Praise for *Uranium*

"In this fine piece of journalism, Zoellner does for uranium what he did for diamonds in *The Heartless Stone*—he delves into the complex science, politics, and history of this radioactive mineral."
— *Publishers Weekly* (starred review)

"Lively, often disturbing history of the largest atom in nature . . . A rich journalistic account." — *Kirkus Reviews*

"*Uranium* is a gripping read—and a terrifying one. [Zoellner] delves into the dark heart of our nuclear ambitions and paints a portrait of a primal force that changed the course of history and threatens our future."
— Charles Seife, author of *Zero*

"This is a wonderful story. Zoellner narrates the history of what uranium has meant to the world over the past century. The book deftly weaves together the fields of mining, science, history, diplomacy, and popular culture."
— James Mann, author of *The Rebellion of Ronald Reagan* and the *New York Times* bestseller *Rise of the Vulcans*

"There may be nothing on this Earth that more powerfully symbolizes human hope and dread than the 'unstable element' known as uranium, and yet, like the thing itself, it lies in the abyss of our imagination. Traveling from the savannas of Congo to the murkier outposts of the former Soviet Union, with a tale that blends suspense, history, horror, and humor, Tom Zoellner has spun the curious, epoch-defining story of this yellow dirt into journalistic gold."
— Tom Vanderbilt, author of the *New York Times* bestseller *Traffic*

Tom Zoellner is the author of *The Heartless Stone: A Journey Through the World of Diamonds, Deceit, and Desire*, named a 2006 Notable Book by the American Library Association. He is the coauthor, with Paul Rusesabagina, of *An Ordinary Man*. He has worked as a contributing editor at *Men's Health* and as a reporter for the *San Francisco Chronicle*.

Uranium

WAR, ENERGY, AND THE ROCK
THAT SHAPED THE WORLD

TOM ZOELLNER

PENGUIN BOOKS

PENGUIN BOOKS
Published by the Penguin Group

Penguin Group (USA) Inc., 375 Hudson Street, New York, New York 10014, U.S.A. • Penguin Group (Canada), 90 Eglinton Avenue East, Suite 700, Toronto, Ontario, Canada M4P 2Y3 (a division of Pearson Penguin Canada Inc.) • Penguin Books Ltd, 80 Strand, London WC2R 0RL, England • Penguin Ireland, 25 St Stephen's Green, Dublin 2, Ireland (a division of Penguin Books Ltd) • Penguin Group (Australia), 250 Camberwell Road, Camberwell, Victoria 3124, Australia (a division of Pearson Australia Group Pty Ltd) • Penguin Books India Pvt Ltd, 11 Community Centre, Panchsheel Park, New Delhi – 110 017, India • Penguin Group (NZ), 67 Apollo Drive, Rosedale, North Shore 0632, New Zealand (a division of Pearson New Zealand Ltd) • Penguin Books (South Africa) (Pty) Ltd, 24 Sturdee Avenue, Rosebank, Johannesburg 2196, South Africa

Penguin Books Ltd, Registered Offices: 80 Strand, London WC2R 0RL, England

First published in the United States of America by Viking Penguin,
a member of Penguin Group (USA) Inc. 2009
Published in Penguin Books 2010

1   3   5   7   9   10   8   6   4   2

Portions of chapter 5 originally appeared in the article "The Uranium Rush,"
by Tom Zoellner, in the Summer 2000 issue of *The American Heritage
of Invention and Technology.*

THE LIBRARY OF CONGRESS HAS CATALOGED THE HARDCOVER EDITION AS FOLLOWS:
Zoellner, Tom.
Uranium : war, energy, and the rock that shaped the world / Tom Zoellner.
p.   cm.
Includes bibliographical references and index.
ISBN 978-0-670-02064-5 (hc.)
ISBN 978-0-14-311672-1 (pbk.)
1. Uranium.   2. Uranium—History.   I. Title.
QD181.U7Z64 2009
546'.431—dc22   2008029023

Printed in the United States of America
Set in Aldus with FFGothic
Designed by Daniel Lagin

CONTENTS

INTRODUCTION

This all began for me at a mesa in Utah called Temple Mountain, so named because its high-pitched walls and jagged spires had reminded early Mormon settlers of a house of worship.

I had driven into the wide canyon at its base, pitched a tent among some junipers, and eaten a can of chili while sitting on a rock and watching the day's last sunlight creeping upward on the salmon-colored walls to the east.

A set of caves, their mouths agape, dotted the face of the cliff. Pyramid-shaped mounds of rock and talus were piled under them, and rotten wooden boards lay half drowned in this debris.

I looked closer and saw that the caves were square, and one appeared to be propped with beams. These weren't caves at all. They were mine entrances.

It now made sense. The valley floor had that ragged and hard-used look common to many other pieces of wilderness in the American West that had been rich in gold or silver in the nineteenth century. A braiding of trails was etched into the dirt, and the slabs of an abandoned stone cabin and shattered lengths of metal pipe were down there, too, now almost obscured in the dusk. The place had been devoured quickly and then spat out, with a midden of antique garbage left behind.

What kind of ore had been carted away from here? Curiosity got the better of me, and I wandered over to a spot down the trail where three other people had also set up camp. They were recent college graduates from Salt Lake City on a spring camping trip. After offering me a beer from their cooler, they told me the holes on the cliff were of much more recent origin than I had thought. Uranium mines had been drilled in southern Utah after World War II, and the mineral had gone into nuclear weapons. This was common knowledge around southern Utah.

Uranium. The name seemed magical, and vaguely unsettling. I remembered the boxy periodic table of the elements, where uranium was signified by the letter *U*. It was fairly high up the scale, meaning there were a lot of small particles called protons clustered in its nucleus. So it was heavy. It was also used to generate nuclear power. I remembered that much from high school science. But it had never quite registered with me that a mineral lying in the crust of the earth—just a special kind of dirt, really—was the home of one of the most violent forces under human control. A paradox there: *from dust to dust.* The earth came seeded with the means of its own destruction, a geologic original sin.

There was something personal here, too. I had grown up in the 1980s in Tucson, Arizona, a city ringed with Titan II missiles. One of those warheads was lodged in a concrete silo and surrounded by a square of barbed wire in the desert about twenty miles north of my high school. It was nearly five hundred times as powerful as the bomb that leveled Hiroshima. Our city was supposed to have been number seven on the Soviet target list, behind Washington, D.C.; the Strategic Air Command headquarters in Omaha, Nebraska; and several other missile fields in the Great Plains. I lived through my adolescence with the understanding that an irreconcilable crisis with Moscow would mean my family and I would be vaporized in white light, and there might be less than ten minutes' warning to say good-bye (the brief window of foreknowledge seemed more terrible than the vaporizing). Like most every other American of that day, I subsumed this possibility and went about my business. There could be no other choice; to dwell on the idea for very long was like looking at the sun.

And now, here I was in a spot that had given up the mineral that had haunted the world for more than half a century. The mouths in the canyon walls at Temple Mountain looked as prosaic as they would have at any other mining operation. They also happened to be in the midst of some of the most gorgeous American landscape I know: the dry and crenulated Colorado Plateau, which spreads across portions of four states in a pinkish-red maze of canyons, sagebrush plains, and crumbling pinnacles that, in places, looks like a Martian vista. This, too, was an intriguing paradox: radioactive treasure in a phantasm landscape. The desert had birthed an awful power.

After my trip, I plunged into the library and wrote an article for a history magazine about the uranium rush of the 1950s, when the government paid out bonuses to ordinary prospectors to comb the deserts for the basic fuel of the nuclear arms race. But my fascination with uranium did not end, even years after that night I slept under the cliff ruins. In the present decade, as the United States has gone to war in Iraq on the premise of keeping uranium out of the wrong hands—and as tensions mount in Iran over that nation's plan to enrich the fatal ore—I realized that I still knew almost nothing about this one entry in the periodic table that had so drastically reordered the global hierarchy after World War II and continued to amplify some of the darker pulls of humanity: greed, vanity, xenophobia, arrogance, and a certain suicidal glee.

I had to relearn some basic matters of science, long forgotten since college. I knew that the nuclear trick comes from the "splitting" of an atom and the consequent release of energy. But why not copper or oxygen or coffee grounds or orange peels or anything else? Why does this feat require a rare version of uranium, known as U-235, that must be distilled, or "enriched," from raw uranium?

I started reading again about the infinitesimally small particles called neutrons and protons packed at the center, or nucleus, of atoms, and the negatively charged particles called electrons that whiz around the nucleus like bees around a hive. Puncture that nucleus, and the electrical energy that bound it together would flash outward in a killing wave. U-235 is uniquely vulnerable to this kind of injury, and I understood this in concept but could not really visualize it until I came across a line written by the physicist Otto Frisch. He described this particular nucleus as a "wobbling, unstable drop ready to divide itself at the slightest provocation." That image finally brought it home: the basic principle of the atomic bomb.

A uranium atom is simply built too large. It is the heaviest element that occurs in nature, with ninety-two protons jammed into its nucleus. This approaches a boundary of physical tolerance. The heart of uranium, its nucleus, is an aching knot held together with electrical coils that are as fragile as sewing thread—more fragile than in any other atom that occurs in nature. Just the pinprick of an invading neutron can rip the whole package apart with hideous force. The subatomic innards of U-235 spray outward like the shards of a grenade; these fragments burst the skins of neighboring uranium nuclei, and the effect blossoms exponentially, shattering a trillion trillion atoms within the space of one orgiastic second. A single atom of uranium is strong enough to twitch a grain of sand. A sphere of it the size of a grapefruit can eliminate a city.

There are other dangers. A uranium atom is so overloaded that it has begun to cast off pieces of itself, as a deluded man might tear off his clothes. In a frenzy to achieve a state of rest, it slings off a missile of two protons and two neutrons at a velocity fast enough to whip around the circumference of the earth in roughly two seconds. This is the simplest form of radioactivity, deadly in high doses. These bullets can tear through living tissue and poke holes in healthy cell tissue, making the tissue vulnerable to genetic errors and cancer.

Losing its center piece by piece, uranium changes shape as it loses its protons—it becomes radium and then radon and then polonium—a lycanthropic cascade that involves thirteen heavy metals before the stuff finally comes to permanent rest as lead. More than 4.5 billion years must pass before half of any given sample decays. Seething anger is locked inside uranium, but the ore is stable and can be picked up and carried around safely as long as its dust is not inhaled. "Hell, I'd shovel some of it into my pillow and sleep on it at night" is a common saying among miners.

Only when the ore has been concentrated to more than 20 percent U-235—which is, thankfully, a job of massive industrial proportions—is there the danger of a spontaneous chain reaction. But after that point, it becomes frighteningly simple. Two lumps of enriched uranium slammed together with great force: This is the crude simplicity of the atomic bomb. (A similar effect can be achieved through the compression of plutonium, a by-product of uranium fission that is covered only briefly in this book.)

Though uranium's lethal powers have been known for less than seventy years, man has been tinkering with it at least since the time of Christ. Traces of it have been found as tinting inside stained-glass mosaics of the Roman Empire. Indians in the American Southwest used the colorful yellow soil as an additive in body paint and religious art. Bohemian peasants found a vein of it in the lower levels of a silver mine at the end of the Dark Ages. They considered it a nuisance and nicknamed it "bad-luck rock," throwing it aside. The waste piles lay there in the forest until the beginning of the twentieth century, when chemists in France and Britain started buying uranium at a deep discount for the first experiments on radioactivity. A West Virginia company briefly used the stuff as a red dye for a line of dishes known as Fiesta Ware. But it was not until the late 1930s when an ominous realization began to dawn among a handful of scientists in European and American universities: that the overburdened nucleus of U-235 was just on the edge of cracking asunder and might be broken with a single neutron.

This was the insight behind America's Manhattan Project, which brought a startling ending to World War II and initiated a new global order in which the hegemony of a nation would be determined, in no small part, by its access to what had been a coloring dye for plates. As it happened, a Japanese company had been among the outfits searching for ceramic glaze at the Temple Mountain site in the years immediately before Pearl Harbor. They left several of their packing crates abandoned in the Utah desert, sun-weathered kanji characters visible on the wood. Had the government in Tokyo understood what really lay there at Temple Mountain, the war might have ended differently.

Uranium did not just reshape the political world. Its first detonation at Hiroshima also tapped deep into the religious part of the human consciousness and gave even those who didn't believe in God a scientific reason to believe that civilization would end with a giant apocalyptic burning, much as the ancient texts had predicted. A nonsupernatural method of self-extinction had finally been discovered. William Butler Yeats was a quarter century ahead of the atomic bomb when he wrote his end-of-the-world poem "The Second Coming" in 1920, but the poem's most famous couplet also happens to be an exact chemical description of the nucleus of uranium:

> *Things fall apart*
> *The center cannot hold*

This unstable element has played many more roles in its brief arc through history, controlling us, to a degree, even as we thought we were in control. It was a searchlight into the inner space of the atom, an inspiration to novelists, a heroic war ender, a prophet of a utopia that never arrived, a polluter, a slow killer, a waster of money, an enabler of failed states, a friend to terrorists, the possible bringer of Armageddon, an excuse for war with Iraq, an incitement for possible war in Iran, and now, too, a possible savior against global warming. Its trajectory has been nothing short of spectacular, luciferous, a Greek drama of the rational age. The mastery and containment of uranium—this Thing we dug up seventy years ago—will almost certainly become one of the defining aspects of twenty-first-century geopolitics. Uranium will always be with us. Once dug up, it can never be reburied.

In this rock we can see the best and the worst of mankind: the capacity for scientific progress and political genius; the capacity for nihilism, exploitation, and terror. We must find a way to make peace with it. Our continuing relationship with uranium, as well as our future as a civilization, will depend on our capacity to resist mirroring that grim and never-ceasing instability that lies within the most powerful tool the earth has to give.

There may be no better place to begin this story than at a different set of ruins. These are in Africa, at the edge of a hole that will not stay closed.

Uranium

1

SCALDING FRUIT

The place is called *Shinkolobwe*.

Its name comes from the Bemba language of south-central Africa and is the word for a thorny fruit resembling an apple, typically cooked by submersion in a pot of boiling water. The outside of the fruit cools quickly, but the inside is like a sponge. It retains hot water for a long time. Squeezing it results in a burn.

The word is also slang for a man who is easygoing on the surface but who becomes angry when provoked.

There used to be a village of the same name at the edge of the pit, but it has since been destroyed by fire. A local story says the area is haunted by a spirit named Madame Kipese, who lives inside the pit. The madame was a cheery and forceful woman when she was alive, but she grew evil after her death and burial. White men came here many years ago to dig the pit and became friendly with her. They may have even had sex with her.

Madame Kipese needs to consume human souls to keep herself strong. She emerges from time to time to kill someone. Unexplained deaths in the area are sometimes attributed to Madame Kipese.

"I would not go there myself," an officer of the federal police told me. He was on the protection staff of Joseph Kabila, the president of the Democratic Republic of the Congo.

"It's a very dangerous place," he went on. "Cell phones burn out when you take them there. Television sets wouldn't work, even if there were a place to plug them in. Be sure you don't wear a T-shirt. You must wear a long-sleeve shirt to protect yourself from the dust. All the men who work there are supposed to wear long-sleeve shirts. Try not to breathe the dust. Whatever you do, don't put any of that stuff in your pocket.

"Are you sure you want to go?"

I told him I was sure.

"You have to cross through at least four roadblocks before you get there," he said. "Each one is more serious. That place is very heavily guarded. It is considered a strategic site. They want to make sure you are not a saboteur. The last line of defense is a squad of United Nations soldiers. I won't be able to help you with them."

I wound up paying him $80 for what he described as a special police authorization.

The next day, I received a photocopy with the presidential letterhead on it. Below the letterhead, in blue ballpoint scrawl, was my name, my passport number, my birthday, and a series of villages I was to pass through on my way. The final destination was marked SHINKOLOBWE.

"That place is highly secure," an employee of a mining company told me. "You're not even allowed to fly over it." He knew this because he had been a passenger in a small airplane the previous year, and the pilot had shown him flight maps. The airspace around the pit was marked RESTRICTED, unlike any other nearby terrain.

Shinkolobwe is now considered an official nonplace. The provincial governor had ordered a squad of soldiers to evacuate the village next to the pit and burn down all the huts in 2004, leaving nothing behind but stumps and garbage. A detachment of army personnel was stationed there to guard the edges and make sure nobody entered.

The government had been embarrassed by a series of accidental deaths inside the pit. Some men were digging inside a jerrybuilt tunnel when it collapsed on them. Eight were killed and thirteen injured.

Fatal accidents are all too common in the illegal mining trade of the Congo. Abandoned mines such as this one are scattered all over the southern savanna, and most of them are still being picked over by local farmers hoping to boost their income by selling a few bags of minerals on the side, usually copper and a smattering of cobalt.

Shinkolobwe was different. This was the pit that, in the 1940s, had yielded most of the uranium for the atomic bombs the United States had dropped on Hiroshima and Nagasaki.

But this was more than a historical curiosity. Shinkolobwe had been a menace for years. The mine shafts were sealed tight with concrete plugs when the Congo became an independent nation more than four decades ago. Yet local miners had been sneaking into the pit to dig out its radioactive contents and sell them on the black market. The birthplace of the first atomic bomb was still bleeding uranium, and nobody was certain where it was going.

Shinkolobwe is in the midst of a pleasant savanna of hills and acacia trees in a region called Katanga, where people have been farming for more than two thousand years with tools made from wood and copper picked from the ground. This place, and the rest of the Congo, had been the private preserve of King Leopold II of Belgium, who claimed the territory for himself when European powers were beginning to plant their flags around Africa in the 1870s.

Leopold had enlisted the help of an adventurer named Henry Morton Stanley, a former staff writer for the *New York Herald*, already famous for his publicity stunt of "finding" the lost missionary Dr. David Livingstone. Stanley set off on a five-year journey to sign land treaties with local chiefs across central Africa, promising liquor, clothing, and some toiletries in exchange for lumber and ivory, in addition to the limitless physical labor of the natives. Within twenty years, Stanley had claimed for King Leopold an estate that was seventy-five times larger that the nation of Belgium. He called it Congo Free State, a corruption of "Kongo," the name of one of the ancient native kingdoms that had signed itself away. Leopold promised to run "this magnificent African cake" for the charity and benefit of the natives.

The Congo instead became a gigantic forced-labor camp. The Africans were threatened with brutal beatings and the amputation of their hands and even beheadings if they failed to collect enough ivory tusks or lumber to satisfy the quotas of Belgian managers. A blanket of rubber trees covered the region, and King Leopold was in an excellent position to fill the demands of the newborn automobile industry, as well as the need for bicycle tires, electrical insulation, telephone wires, gaskets, and hoses. By the turn of the century, more than six thousand tons of rubber sap was leaving the Congo, all of it tapped by Africans threatened with beatings, imprisonment, kidnapping, murder, and systematic rape. Those deemed lazy had their hands and forearms hacked off by members of Leopold's security organization, the Force Publique, who sometimes collected baskets of severed hands to prove to their supervisors they had been diligent in encouraging the harvest.

As stories of the abuses leaked out, the Congo became a symbol of greed. Among Leopold's many critics was Joseph Conrad, who had taken a job on a steamship responsible for moving a load of railroad ties up the Congo River. What he saw there disgusted him. In his novel *Heart of Darkness*, he wrote of a rapacious company modeled after some of Leopold's concessionaries. "To tear treasure out of the bowels of the earth was their desire," he wrote, "with no more moral purpose at the back of it than burglars breaking into a safe."

Leopold died of an intestinal blockage in 1909, having extracted from the Congo a personal fortune exceeding $1 billion. He never once visited it. The government of Belgium took over the estates and was only slightly more merciful than the king, moderating but not ending the reign of the Force Publique and preserving the system of forced labor under the rule of monopoly companies. The largest was a mining giant called Union Minière du Haut Katanga, which started exploiting copper in the southern tail section. Leopold hadn't much cared about mining (it was too expensive; rubber was much easier), but the company discovered what he had missed: generous quantities of bismuth, cobalt, tin, and zinc at shallow depths. Under a rug of grass, a golden floor.

Delighted executives called it *un scandale géologique*—a "geological scandal"—and built an empire of mills, furnaces, and rails in the bush. Locals were paid the equivalent of 20 cents a day to break rocks and push carts. It amounted to a version of debt slavery: Taxes were kept purposefully high, and workers were not permitted to select their own occupations. The men slept eight to a hut in settlements ringed with barbed wire to prevent them from leaving before their contracts were up. Typhoid and dysentery were rampant, and about one miner out of every ten died every year from disease, malnutrition, rock collapses, or beatings administered by the Belgian managers. "The food of the workers is awful," reported one observer. "They are only fed during the week with flour or corn."

One of these sites had been Shinkolobwe, where patches of high-grade uranium had been found in 1915. Radium was the most valuable substance on earth at that time; American doctors were calling it a miracle cure for cancer, and some were counseling their patients to drink a weak radium solution sold under the name Liquid Sunshine. A gram of it could fetch $175,000, thirty thousand times the price of gold.

Union Minière tore apart the hill and started tunneling underground, forcing more than a thousand African laborers to dig into what would turn out to be the purest bubble of uranium ore ever found on the earth. The workers were made to carry sacks of the velvety black stone more than twenty kilometers to the railhead, where the sacks were sent to port and then shipped by ocean steamer to Belgium. The uranium-rich leftovers, known as tailings, were simply thrown away. Uranium was interesting only because it hosted tiny bits of radium. By itself, it was considered worthless: a trash rock.

When the Nazis invaded Belgium in 1940, Union Minière moved its headquarters to New York. War was lucrative, and the United States would soon become the world's largest user of Congo cobalt, an important metal for the manufacture of aircraft engines. American consumption would rise by a factor of ten before the end of the war, and the Congo mines started operating around the clock.

But there was something of much greater value than copper that the United States would need. On the afternoon of September 18, 1942, a U.S. Army colonel named Kenneth D. Nichols paid a visit to Union Minière. He wore a coat and tie for the occasion. Nichols had just been hired to help administer the Manhattan Project, the top-secret effort to build the atomic bomb, and he was there to buy the waste uranium from Shinkolobwe, which was one of the only known sources of the mineral.

Nichols left the office thirty minutes later with some figures on yellow scratch paper that formed the basis of a secret contract between the United States and Union Minière. The mine would go on to supply nearly two-thirds of the uranium used in the bomb dropped over Hiroshima, and much of the related product plutonium that went into the bomb dropped on Nagasaki. The pit was deepened and widened, and the ebony vein of uranium would go on to feed the massive American buildup of nuclear weapons after the war.

"A freak occurrence in nature," Nichols called it. "Nothing like it has ever again been found."

For the next two decades, Shinkolobwe enjoyed a mystique as the most famous source of the most powerful substance on earth. Access to the site was forbidden, and the closest a visitor could get was to view the giant block of pure uranium the company put on display in the nearby city of Elisabethville. "As big as a pig, its color was black and gold, and it looked as if it were covered with a green scum," reported one observer. Visitors were warned not to get too close with their cameras, lest their film be fogged and ruined. A sign said: ATTENTION. BLOC RADIOACTIF!

The Belgians had expected to rule their colony for more than a century, but increasing violence in the capital convinced them to step aside and grant the Congo its independence in 1960. They left behind not so much a country as a plantation. There were only seventeen university graduates left to run a new nation of sixteen million people. With American backing, a twenty-nine-year-old army officer named Joseph-Désiré Mobutu seized power in a coup and would, over time, set himself up as a secular messiah even as he looted the nation as systematically as had King Leopold. The currency, the major river, and the entire country itself were renamed Zaire. New parents were discouraged from giving their children European names, with the president setting the example by renaming himself Mobutu Sese Seko Kuku Ngbendu Wa Za Banga, which officially meant the "Warrior who knows no defeat because of his endurance and inflexible will and is all powerful, leaving fire in his wake as he goes from conquest to conquest." (A translation in a related dialect is "Fierce warrior and a cock who jumps on any chicken that moves.") He took a cut from virtually every business in his country, siphoning off $4 billion and building luxurious marble palaces for himself all over his wretchedly poor nation. The once-promising economy went into a tailspin. Roads fell apart. Farms dehydrated in the equatorial sun. Union Minière's property was nationalized and then looted.

But not Shinkolobwe. The ore was already running low when the colonial era came to an end in 1960, but the managers feared that such a lethal substance would fall into the wrong hands. They poured concrete into the shafts and carted off the equipment. Scavengers tore the metal from the uranium warehouses. The workers' village was evacuated and sealed off, and weeds began to sprout inside the shells of brick town houses. Mango trees drooped, nodded, and eventually toppled onto the deserted streets. Shinkolobwe crept back, day after day, into a state of nature.

In the confusion of the Belgian retreat from the Congo, the CIA's station chief, Larry Devlin, received an unusual cable from his bosses. Could he go out to the campus of the University of Léopoldville and take the uranium fuel rods out of the nuclear reactor? The CIA further instructed him to find a deserted spot in the African countryside to bury the rods until the rioting calmed down.

Devlin was at a loss. He had no training for handling hot radioactive goods. And as for sneaking out to the jungle to bury them, "I could not think of a way to do that in a country where a white man stood out like a cigar store Indian," Devlin recalled in his memoirs.

This reactor owed its existence to Shinkolobwe. A priest named Luc Guillon, the founder of the Congo's only university, had argued that the colony had done a patriotic service by allowing its uranium to be acquired by the Manhattan Project, and therefore deserved its own piece of the nuclear future. Guillon was allowed to buy an experimental reactor from the U.S. company General Atomics and install it on the edge of campus. Africa's first reactor went critical in 1959, to local fanfare. But one year later, the Congo was a newborn independent nation in a state of war, with uranium fissioning behind a flimsy fence while gun battles flared outside.

Devlin sought gamely to carry out his order. He drove out to the university through uneasy streets, passing three roadblocks on the way, and explained his mission to Guillon, who told him that hiding the uranium in the jungle was "a crazy idea." The rods were safer just left in place. And there they remained.

Mobutu loved the reactor—it was a source of national prestige—and he made sure to attend all its ceremonial functions. But the facility grew shabbier with each passing year as the economy decayed. The cooling water is now said to be filthy, and the power is switched on only once a week, to ensure that it still functions. At only one megawatt, the plant is a toy by world standards (by contrast, the Indian Point station outside New York City has a capacity of two thousand megawatts), but its presence in such an unstable country has long been a worry. A meltdown would not destroy the city; the facility is too small for that. But an accident or sabotage could kill hundreds of people and leave the neighborhood toxic for decades. Guillon's sense of topography was also poor. The reactor is on sandy soil about a hundred yards from a hill that tends to crumble and slide in the rain.

Security here has been a long-standing joke. When the British journalist Michela Wrong visited the facility, she found "no carefully monitored perimeter fences, guard dogs, or electric warning systems. Only a small sign—one of those electrons-buzzing-around-an-atomic-core logos that once looked so modern and now look so dated—alerts you to the presence of radioactive material."

The facility is guarded by a low fence sealed with a padlock. Two of the uranium rods in the facility were stolen in the 1970s without anybody realizing they had disappeared. One of them eventually turned up in Rome in the possession of members of a Mafia family, who were offering it for sale to "Middle Eastern buyers" who turned out to be undercover Italian police officers. Only at this point was the long-ago burglary discovered. The other rod has never been found.

For the Congolese, this was just another sour joke; another application of the catchall term *Article Fifteen*, which is a supposed unwritten clause to the constitution that allows a certain amount of dishonesty in one's personal affairs.

Article Fifteen is how the Congolese squeeze out a living by any method possible: by selling cigarettes in the nightclubs after work, by smuggling copper into neighboring Zambia, by printing up school diplomas with a forged signature, by renting out the boss's car as a taxi before he finds out. The phrase, according to Wrong, seems to have originated in the early 1960s when the leader of a breakaway republic in south Kasai grew weary of the pleas for shelter and food from the refugees flooding into his region. He gave them this imperial retort: "It's your country, so fend for yourself." This statement was repeated so often, and with such amusement, that some said it ought to be posted as a motto in government buildings. In the wreckage of an economy that is the Congo, this is how things get done.

Mobutu was overthrown by a rebel army in 1998, shortly before he died of prostate cancer. The nation was renamed the Democratic Republic of the Congo, and the flag of Zaire was replaced with a blue banner with a gold star in the middle—a flag that bore a strange resemblance to King Leopold's flag. The new government promised better roads and schools. But the culture of graft known as Mobutuism is still pervasive. Diplomas and government contracts are still for sale to those who ask. A request for *sucre*, or "a sugared drink," is the usual euphemism for a bribe.

Occasionally the request can be blatant. I once had a brief and inconsequential conversation with one of the top officials at the Ministry of Mines in a southern city. I had met him in his crumbling office building, where the hallways were dark and the lights were broken. When I made motions to leave, he said, out of the blue, "So, don't you have something there in your pocket for me?"

This was really nothing out of the ordinary: only a customary application of Article Fifteen. In a country raped so long and so badly by those who sought its riches, this is the way many official things are done. It is also the governing principle in the uranium ruins.

Shinkolobwe does not appear on most local maps, but is not a difficult place to find. I hired a translator named Serge, and we rented a Toyota Land Cruiser for a few hundred dollars in the city of Lubumbashi, once called Elisabethville, the principal railhead for most of the ore trains that used to run to the Atlantic. The city's weak economy still thrives on minerals, both legal and bootleg. Chinese companies have established a presence as buyers of the copper and cobalt picked out of the open pits.

We left the city at dawn and headed north on a potholed national highway that faded into dirt, through forests of eucalyptus and acacia. Ore furnaces lined the road near the town of Likasi, and the road was dusted black with cobalt. Serge turned onto a rutted sidetrack in the hilly country north of Likasi, and we soon got bogged down in the mud. He gunned the engine while I got behind the Land Cruiser and pushed. A group of local farmers happened down the road at that moment, all of them wearing T-shirts and carrying machetes. They joined me in the pushing.

One of them was a man in his thirties with calloused hands and a red jersey. He told us his name was Alphonse Ngoy Somwe and that he had worked as a miner at Shinkolobwe, where copper was usually the big thing. There had been at least one time, however, when he had looked for uranium.

A few years ago, he recalled, some white men had come to buy their ore and had waved electronic devices over the rocks. This would not have been unusual in itself, as the cobalt ore is sometimes vetted for radioactivity, but the men seemed to be looking for uranium specifically. This surprised Somwe. It had previously been considered garbage, a nuisance.

He said he didn't want to do mining anymore—"it kills"—but after we pushed the Toyota loose, he agreed to show us the way.

We bounced past a Pentecostal church made out of poles and grass and, shortly thereafter, came to a spot where the road took a plunge into a rocky valley, too precipitous for the Land Cruiser to handle. Somwe told us the mine was about four miles farther. Serge pulled the vehicle off to the side. I shouldered my pack, and we all started walking.

A substantial amount of uranium has been smuggled out of the Congo in the last decade, and the source is almost certainly Shinkolobwe.

In October 2005, a customs official in Tanzania made a routine inspection of a long-haul truck carrying several barrels labeled columbite-tantalite, otherwise known as coltan, a rare metal used in the manufacture of laptop computers and cell phones. But he found a load of unfamiliar black grit instead.

One of his bosses later recalled the scene to a reporter: "There were several containers due to be shipped, and they were all routinely scanned with a Geiger counter. This one was very radioactive. When we opened the container, it was full of drums of coltan. Each drum contains about fifty kilograms of ore. When the first and second rows were removed, the ones after that were found to be drums of uranium."

The truck had come into Tanzania from neighboring Zambia, but had started its journey in the Congo. This was an echo of an incident three years prior in the town of Dodoma when a large cache of raw uranium, sealed in plastic containers, was confiscated. A United Nations panel came in to investigate and concluded the source of both shipments had been illegal mining at Shinkolobwe.

"The frequency of seized consignments in the Central African region leaves no doubt that the extraction and smuggling must be the result of organized efforts, and that these illegal activities must be highly rewarding financially," said their final report. At least fifty cases of uranium and other radioactive material had been confiscated around the capital city of Kinshasa in the last eight years and even more was going undetected. "Such incidents are far more frequent than assumed," said the inspectors.

The clandestine picking of uranium is not hard to conceal in the midst of so much other petty corruption. In most of Union Minière's abandoned pits, there is an active hunt for what is called "Congo caviar"—the rich mineral blend of cobalt and copper harvested by scavengers and purchased by speculators. This activity is supposed to be illegal, but it has been widely tolerated for more than a decade. The miners work in T-shirts and flip-flops and dig out the chunks of "caviar" with shovels, picks, and their bare hands. Approximately fifty thousand to seventy thousand people are doing this on any given day.

The work is dark and dangerous. The miners sink handmade shafts that go perhaps forty feet down, then kink crazily in all directions. The horizontal chambers are known as galleries. They are no larger than crawl spaces; there is barely enough room to make a half swing with a pick. At least forty people a year are killed in tunnel collapses. There is little chance of underground rescue; the galleries become tombs. Giant fissures have appeared on the floor of some pits, indicating that the honeycomb of tunnels underneath has weakened the ground to the point of fracture and collapse.

Once mined, the caviar is packed into threaded plastic bags that resemble sacks of corn or wheat and sold to brokers called *négociants*, who turn around and sell to a trading company. The cobalt is particularly prized and fetches high prices. It is a vital metal in the construction of jet engines and turbines. Energy-hungry China is a primary buyer. But in the majority of cases the minerals leave the country illegally, without being recorded and without being taxed. The usual route is through Zambia. And at every step in this unofficial process, from the mine to the border, successive layers of police and inspectors demand a cut.

"Those who work in the sector have little choice in the matter; their ability to work, to buy and sell is dependent on paying these bribes," reported the British advocacy group Global Witness. "The practice has become so institutionalized that it is no longer challenged."

A freelance miner named Bedoin Numbei, whose T-shirt bore the legend ALADDIN, LAS VEGAS, told me this was true. He himself had snuck into the mine at Shinkolobwe to mine ore with the approval of the same guards who were supposed to be preventing this activity. "You just have to pay a little gift to the soldiers and you can go in there at night," he told me.

A common joke among nuclear policy analysts is that the best way to move an atomic bomb across a national border is to hide it inside a truckload of marijuana. In other words, smuggling routes used by average criminals provide good cover for the occasional piece of nuclear merchandise. This appears to have been the case at Shinkolobwe. A dossier from the government in Kinshasa reported that radioactive products, with no weights reported, have been sold in Katanga at prices ranging from $300 to $500 to a variety of traffickers from India, China, and Lebanon. Article Fifteen had been applied to more than just cigarettes, gasoline, and batteries. Uranium ore was now for sale.

After about two hours, we came to the remains of a metal fence nearly covered in the jacaranda trees on the side of the road.

"This was the beginning of the secure zone," said Serge. He pointed out a small concrete foundation off to one side, also hidden in the brush. There were a few bricks scattered about. It appeared to have been a guardhouse.

Somwe took us down a winding path through man-high grass that eventually led into the ruins of a European-style village in a clearing among mango trees. We walked down a narrow lane that separated two rows of town houses. The walls had mostly collapsed, and those that stood upright were speckled with dark moss. Grass obscured the floors. A line of what had been streetlights was now just hollow steel stumps; the streetlights had been cut down like cornstalks. Mud huts of more recent construction were off to the side, their roofs missing. It felt as if we were walking through the leavings of a bygone civilization—a garrison on the Roman frontier, perhaps, or one of the forgotten silver villas in the Andes. But this was antiquity of the atomic age.

Somwe motioned us onward. We walked about a half mile down a concrete road, past mounds of black dirt, old slag. This was the outer fringe of what had been the B Zone, the heart of the uranium mining operation run by the Belgians fifty years ago. Shards of iron pipe and green chips of oxidizing copper lay scattered on the path. To the west were the metal skeletons of what had been a warehouse and a water tower.

When we passed through a gap in the trees, a panorama suddenly opened. Across a wide clearing in the forest, it was possible to see a line of trees a mile away, across a low man-made canyon whose sides were stained black and brown and whose bottom held pools of cloudy green water. This was the Shinkolobwe pit, the womb of the atomic bomb. On a different side of the world, a quarter million Japanese had been killed with the material from this cavity.

Clinging to the edge of the pit was a steel shaft. It was crowned with a slab of concrete, which gave it the appearance of a toadstool. This was one of the entrances that Union Minière had plugged in 1960, in an attempt to keep anyone from getting at the ore remaining inside. The shaft went almost six hundred feet down. There were no freelance miners anywhere in sight, but the soil in the center of the pit had been thoroughly worked over. Broken wood slats were littered about the slopes, the remnants of jerrybuilt mine works.

The three of us stood at the edge of the pit and looked in for a while. None of us spoke. A few fat cumulus clouds drifted overhead.

We were there for several minutes before I realized that I still had the "letter of authorization" from the police official in my backpack. I hadn't needed to withdraw it because we hadn't encountered a single roadblock. Nobody was guarding Shinkolobwe. We had walked right in.

2

BEGINNINGS

The story of the atomic bomb began in the Middle Ages, in a forest surrounded by mountains.

The range was known as Krušnè Hory, or the "Cruel Mountains," because of the harsh winter winds and snows near the summits. They took on an even more melancholy appearance in the spring, when creeks drained snowmelt from the meadows and fog pooled in the valleys. Only bears, wolves, and a few tough hermits could live here. Because of the mountains' obscurity, they became a hiding place for refugees during the religious wars of the fifteenth century, when the reformer Jan Hus was burned at the stake in Prague and a radical sect of his followers, known as Taborites, started slaughtering their neighbors and then retreating into walled towns to wait for the end of the world.

Silver was discovered in a creek on the southern slope in the 1490s, which changed everything. Restless young men from farm villages flooded in to comb the forests for easy money; silver was said to be so plentiful that crumbs of it could be seen clinging to the roots of upended trees. When the surface ore ran out, the migrants started hacking into the slopes. They built rude cabins on the hillsides and smelters to roast the ore. The Czech historian Zbynek Zeman cites a pioneer song from the Cruel Mountains that captures the mad glee of that era:

> *Into the valley*
> *Into the valley*
> *With mothers*
> *With all!*

One local strongman, a count named Stephan Schlick, took over the valley in 1516 and started cleaning up the mess. He hired some journeymen to stabilize the mine shafts and brought law and order to the ramshackle outpost that had taken root about halfway to the summit. Seeking a bit of class, he called his roaring camp St. Joachimsthal—or "Saint Joachim's Valley"—after the father of the Virgin Mary.

The town quickly became the third most populous in all of Bohemia. Taverns sprang up on the valley bottom, where fog and hearth smoke and gases from the smelters thickened the air on cold days. Chicory stew and potatoes were the usual suppers. An early resident complained of "tricksters, riffraffs, and low-lives" as well as "lazy craftsmen, for whom the room and the stool were too hot."

The silver in the valley made it an inviting military target, and so Count Schlick invested heavily in fortifications. On a promontory overlooking the valley, he built a stone castle with a deep cellar and told his metallurgists to start minting coins inside. The first silver disks they produced featured an engraving of the Bohemian king Ludwig I over the name of the town. More than that, they were *big*—larger and weightier than any other coin in circulation. Carrying one in a pocket made a person feel instantly rich. They became a regional sensation.

Count Schlick disappeared after marching off to fight the Turks in 1528, and his mines were eventually annexed by the Hapsburg house of Vienna, which ensured a wider reach for the valley's silver (and handsome seigniorage for the royal sponsors). The big, heavy coins became a staple in market tills and court treasuries in France, Spain, and England. It was a publicity coup for the valley. Merchants began calling the coins Joachimsthalers, later shortened to "thalers," which became bastardized to "dollars" in English-speaking regions.

In this way, the U.S. dollar took its linguistic roots from the mine shafts of St. Joachimsthal, which, in addition to a river of silver, yielded a curious material that stuck to the miners' picks. Dark and greasy, it typically showed up in kidney-shaped blobs, with the neighboring rocks stained brilliant shades of green, orange, or yellow.

The miners nicknamed the stuff *"pechblende"* (the German word *blende* means "mineral," while *pech* can mean both "tar" and "misfortune"; it was literally the "bad-luck rock") and tossed it aside. Seeing this *pechblende*—the English word was *pitchblende*—was never welcome: It usually meant a particular vein of silver had been cleaned away, leaving nothing but mineral garbage, and the miners would have to endure the backbreaking chore of sinking another shaft.

When the silver ran out, the town nearly died. An epidemic of bubonic plague arrived in 1613 and an invading Swedish army sacked the town, reducing it to a valley of burned stumps. The watchtower stood half ruined. Crop failures had forced many to eat boiled hay and insects. Some of those who remained were also stricken by a mysterious disease called *bergkrankheit*, or "mountain disease," which had started approximately fifteen years after the first shafts were dug. Nobody knew what caused it, though arsenic was suspected. Hundreds of people came down with a persistent hacking cough and spit up blood. Death arrived after a few pained months. The disease did not seem to be linked to the plague or to other common maladies of the lungs, but local physicians were at a loss as to how to treat or explain it. "Their lungs rot away," reported Georgius Agricola, who theorized it was due to "pestilential air" in the shafts. But nobody thought to connect it to the velvety black rock.

More than a century later, a sample found its way to a thirty-seven-year-old Berlin pharmacist named Martin Klaproth, who had first studied to be a priest but taught himself chemistry while working as a clerk. He had already gained a small measure of local fame for exposing a scam against Empress Catherine II, who had paid for a remedy called nerve drops. Klaproth proved the drops were nothing more than a mixture of iron chloride and ether, and won the court's gratitude.

In the spring of 1789, he examined the waste product from St. Joachimsthal and realized that, whatever the stuff was, it was associated with lead. When he heated it in solution, it produced a type of yellow crystal the pharmacist had never seen before. Klaproth added wax and a little oil to isolate a heavy grayish residue that he called "a new element which I see as a strange kind of half-metal." Very strange, in fact: It created vibrant yellows and greens when added to glass.

Klaproth refused to name this new coloring agent after himself, as would have been the custom. He instead gave the honor to a new planet in the sky, Uranus, which had recently been discovered by an amateur astronomer in Britain. The new metal was called "uranium" until a more suitable moniker could be found. But none ever was.

The pharmacist had, indirectly, given the metal the name of the Hellenic sky god Uranus. According to the Greek creation story, Uranus had visited the earth every night to make love with the ground and bring forth children who would one day grow into the mutated Cyclops and the Titans. Uranus hated his own children and ordered them chained in a prison deep inside his wife, the earth. One of the most violent of his children rose up from his prison, castrated his father, Uranus, and tossed the severed penis and testicles into the sea. These organs grew into avenging spirits called Erinyes, or the Furies, who occasionally returned to earth for the persecution and damnation of men who upset the natural order.

One of the first people to see the danger of this new substance was not a scientist himself. He was instead a writer of science fiction.

Herbert George Wells was a schoolteacher and a drape hanger from a small town in Kent who found time in 1896 to write *The Time Machine*, a book that would make him famous. He wrote at a breakneck rate, turning out articles and books concerned with socialism, sexuality, violence, evolution, and, above all, man's ability to claw his way upward with logic and technology. His novels, which he called "scientific romances," echoed his politics. *The War of the Worlds, The Island of Dr. Moreau, The First Men in the Moon,* and *The Invisible Man* invoked fantastical or warped versions of the future to illuminate home truths about mankind.

As the likelihood of war fell over Europe in 1914, H. G. Wells retreated to a chateau in Switzerland and dashed off an antiwar novella he called *The World Set Free*. It is not so much a coherent story as it is a jumble of Wells's political ideas, voiced by dull characters who are abandoned shortly after they are introduced. The plot spans thousands of years and is bound together with a single thread: an element called Carolinum, a fictional stand-in for uranium.

Wells somehow managed to make this mineral the only interesting character in his entire novella. Unstable at the core and casting off tiny bits of itself with each passing second, it first excites the imagination of a chemistry professor. "A little while ago, we thought of the atoms as we thought of bricks," he tells his class, "as solid building material, as substantial matter, as vast masses of lifeless stuff, and behold! Those bricks are boxes, treasure boxes, boxes full of the intensest force!"

Before long, this secret is in the hands of scientists who, with deadly imagination, start making "atomic bombs" (Wells appears to be the first person in history to use this phrase). Two sets of allies, the Free Nations and the Central Powers, soon wage a nuclear war and turn each other's capitals into lakes of flame. The heroic King Egbert rallies a council and resolves to safeguard the entire planet's reserves of Carolinum. Whoever could control this rare metal, he realizes, could control the world. After escaping an assassination plot involving an atomic bomb planted in a hay lorry, King Egbert ushers in a paradise on earth, with the fatal element under permanent lock and key. Man had faced down a mineral demon.

Wells was a literary star of his day, but his novella sold poorly and was dismissed by the critics. The *Times* of London derided it as "a porridge composed of Mr. Wells' vivid imagination, his discontents and his utopian aspirations."

But at least one part of the narrative was faithful to reality: Wells had managed to write an accurate physical description of the faux uranium.

He had become fascinated with the emerging field of atomic physics after reading a copy of an academic treatise called *The Interpretation of Radium*, written by Frederick Soddy, a talented chemist from Cambridge, who had helped investigate the decay of thorium, uranium, and radium and concluded they were casting off tiny fragments he named alpha particles.

This disintegration did not seem to be occurring in the molecules, but rather *inside* the atom. Soddy estimated that the energy there must be enormous, perhaps as much as a million times greater than that of any other molecular change. This shakiness at the core of these atoms appeared to be so potent that, for Soddy, conventional physics could not make sense of it. In a speech to the Corps of Royal Engineers in 1904, he mused that a man who truly comprehended uranium could build a weapon that would destroy the earth. In a less belligerent vein, he predicted that a ton could light London's lamps for a full year.

This speculation was possible only because of a lucky accident a few years before. A French chemist named Henri Becquerel heard reports of emissions from cathode tubes that had been nicknamed "X-rays," the X being a placeholder for the unknown source of energy. Becquerel thought he might be able to solve the mystery by experimenting with various types of fluorescing substances. He sprinkled a little of the compound onto a photographic plate and exposed it to the sun outside the laboratory window.

When he tried a new type of salt—uranium potassium sulfate—a silhouette appeared on his plate. This was no surprise: Uranium was already known to fog photographs. Becquerel concluded that sunlight had triggered some kind of gaseous emission. The last days of February 1896 were gloomy and overcast in Paris, and Becquerel decided to suspend his experiments, leaving the plates carefully salted with uranium inside a drawer to await his return. When he did, he found a surprise. The photographic plates showed the same patterns as before. The fogging continued even in the dark—sunlight therefore had nothing to do with it. Whatever this was, it was no gas. This was a constant source of energy, indifferent to its environment. Its luminosity came from within.

What Becquerel was seeing, of course, was evidence of radioactivity—the tiny particles that uranium is always casting away from itself.

He could not have known it at the time, but this instability that characterizes uranium was due to its heaviness. With ninety-two protons jammed together in its nucleus, uranium is the fattest* atom that occurs in nature and is therefore in a constant state of disintegration.

A useful metaphor for radioactivity might be found in architecture, and a good place to look is at a curious building on West Jackson Boulevard in downtown Chicago. The Monadnock Building is sixteen stories tall, and when completed in 1891, it was an object of popular marvel

---

* At least twenty-six elements with nuclei that exceed ninety-two protons, known as the *transuranics*, have since been created in laboratory settings, but—other than trace amounts of plutonium and neptunium—none of them appear in nature, and they must be made artificially.

because it was the tallest building in the world. Steel was still expensive, and architects had not yet learned how to build with the kind of internal frames that could lift an edifice a hundred stories or more. All the weight of an edifice, therefore, had to rest on its walls, as it had since before the time of the Bible. The Monadnock was stone and mortar, and sixteen stories was the breaking point with those materials. Any higher and the whole thing would fall into a pile of rubble, or require walls so big and windows so small that the rooms would have resembled dungeon cells. Even so, the walls of the Monadnock are grotesquely thick, bulging six feet outward at the ground level. The building is so obese with masonry that it sank nearly two feet into Chicago's lakefront soil after it opened. It is still the tallest building in the world without a steel frame, and it represents a monument of sorts: the very brink of physical possibility, like the notion of absolute zero, at 459 degrees below Fahrenheit, beyond which molecules stop moving altogether and cold can get no colder.

There is a similar invisible limitation inside atoms, and uranium is the groaning stone skyscraper among them, pushing the limits of what the universe can tolerate and tossing away its bricks in order to forestall a total collapse. This is radioactivity.

Becquerel's discovery attracted the attention of another Parisian—a thirty-one-year-old graduate student named Marie Sklodowska who had recently emigrated from Poland and married fellow physicist Pierre Curie. While working on her doctoral thesis, she suspected there must be traces of an unknown element, spraying even more radioactivity, hiding deep within uranium. Without having seen it, and acting only on a hunch, she named it radium.

Pierre and Marie Curie wrote to the Austrian Academy of Sciences to ask about the slag heaps at St. Joachimsthal. At the turn of the twentieth century, Martin Klaproth's "strange new metal" was still regarded as a worthless tagalong of silver, good only for making colorful stains for ceramics. The leftover piles of "bad-luck rock" had simply been dumped in the pine forest.

One ton was released to the Curies free of charge, and the remaining stocks were priced at a deep discount. In the summer of 1898, a horse-drawn wagon delivered several canvas bags full of sandy Bohemian pitchblende. Mingled inside were stray pine needles from the trees. Pierre had secured the use of a shed that had previously been used to dissect human corpses. The only furniture inside was a cooking kiln and a series of pine tables where Marie piled her pitchblende.

When dissolved in solution, boiled, and then cooled, the pitchblende formed crystals in much the same way that cooling saltwater leaves flakes of salt on the edge of a glass. "Sometimes I had to spend a whole day stirring a boiling mass with a heavy iron rod nearly as big as myself," she recalled later. Marie examined the crystals with an electrometer, setting aside the specks with the most powerful radioactive signatures.

She was eventually able to isolate a tenth of a gram of radium chloride and prove it was a new element that deserved its own spot in the periodic table. Marie and Pierre were jointly awarded the 1903 Nobel Prize in physics, shared with Becquerel, for their discovery of radiation phenomena. The French newspapers fell in love with the Curies and their cadaver shed. Pierre made a show of exposing his arm to radium to create a burn, which healed after two months, to demonstrate what he called radium's abilities to cure cancer.

Doctors confirmed that he was right. Concentrated doses of radium could indeed shrink and even eliminate tumors. The radiation seemed to kill younger cells—in particular, the cancerous ones—while leaving healthily matured cells untouched. "It was just as miraculous as if we had put our hands over the part and said, 'Be well,'" reported one Johns Hopkins physician after giving radium treatments to a man with a bulbous tumor on his head. The tumor had vanished after radium treatments. *Cosmopolitan* magazine trumpeted radium's virtues in an article that called it "life, energy, immortal warmth" and "dust from the master's workshop." A San Francisco company added trace amounts to chocolate bars. Glow-in-the-dark crucifixes were coated with radium paint. A potion called Radium Eclipse Sprayer claimed to work as both a bug killer and a furniture polish.

The St. Joachimsthal mine directors, who had been happy to give away their trash for free when the Curies had asked for it, now set to exploiting the "bad-luck rock" as the centerpiece of a spa business. They built a two-mile pipeline to carry hot water to the center of the medieval town, which was close enough to Vienna and Prague to attract some of those cities' smart sets. Coach tours were commissioned, and a rail spur was added. St. Joachimsthal was soon welcoming twenty-five hundred visitors a year. One of them was a blue-eyed American private school student named J. Robert Oppenheimer, who would later say his interest in science began when his uncle gifted him a collection of colorful stones picked from the St. Joachimsthal mines.

New brick town houses sprang up in place of the cottages where miners had died of the mysterious wasting disease. The Radium Palace Hotel, with a grand marble staircase and fountain garden in the front, was built on a slope overlooking the valley. A local brewery turned out bottles of Radium Beer. Marie Curie herself was invited to make a sentimental pilgrimage to the "birthplace" of the mineral that had made her and her husband famous.

But, in fact, both Marie and Pierre were ill with radiation sickness, unknown at the time. When the physicist Ernest Rutherford paid the couple a visit in Paris in 1903, he noticed that Pierre's fingers were red and inflamed, shaking so badly he could barely hold a tube of radium salts that he was showing to his guests. Pierre was too sick to present his Nobel lecture and had to postpone it for two years. When he finally stood in Stockholm to receive the award, his tone was hesitant.

"Is it right to probe so deeply into nature's secrets?" he wondered. "The question must here be raised whether it will benefit mankind, or whether the knowledge will be harmful. Radium could be very dangerous in criminal hands."

He finished his address on a hopeful note, invoking the name of the Swedish chemist Alfred Nobel, who had become rich from inventing dynamite and was savaged as a "merchant of death" by the newspapers, but who had also become a pacifist and endowed the famous prizes that celebrated advances in science, art, and peace.

"The example of the discoveries of Nobel is characteristic, as powerful explosives have enabled men to do wonderful work," concluded Pierre Curie. "They are also a terrible means of destruction in the hands of great criminals who are leading the people toward war. I am one of those who believe with Nobel that mankind would derive more harm than good from these new discoveries." Scientific curiosity was moving forward, a relentless force.

This strange energy inside uranium, this radioactivity, was causing scientists to reexamine some long-standing assumptions about the cosmos. The basic understanding of the atom, for example, was about to take a radical shift.

An atom is approximately one hundred millionth of an inch across, tiny enough to be everywhere and invisible at the same time. The existence of these universal building blocks was first theorized in the fourth century B.C. by a wealthy Greek dilettante named Democritus, who was amplifying an earlier theory from a philosopher named Leucippus. Both men were in rebellion against the concept of monism, which holds that all substance, including empty space, is a single unified object bound together with invisible connections. Democritus proposed the concept of a pixilated world made up of tiny basic balls of matter that were impossible to split. He coined the word *atom*, which literally means "indivisible," and described atoms as being constantly vibrating and in motion, banging against one another and binding to one another in distinct patterns to form the minerals and vegetables of the world: gold, sand, trees, the ocean, even man himself. One of his followers compared the movement of atoms to the lazy meanderings of dust particles inside a sunbeam.

The exact nature of these atoms was left murky, but Democritus speculated that they had the reduced characteristics of their grander forms: Atoms of water were slippery, atoms of sand were sharp and jagged, atoms of fire were hot and red, and so on. He even believed the human soul was made of atoms, which disperse to the winds upon death, forever obliterating that person. Democritus was not a believer in immortality.

The notion of a miniature world wriggling out of the sight of mankind was disturbing to more classic theorists, including Plato, who preferred to think of the inner firmament of the world as a divination of the gods, and not anything that could be expressed as tiny dots. The atomic proposition also seemed to wreck the famous paradox of Zeno's arrow, which must travel half the distance to its target, but before that, half of that half, and before that, half of the half of the half, and so on into infinity, meaning that the arrow ought to be in flight forever, but it clearly was not. Space and matter were not, therefore, infinitely divisible. They were *hardened* on some level, just as Democritus was suggesting. Plato was supposed to have told his friends he wished all sixty books by Democritus could be burned.

The wish might as well have come true, for none of Democritus's writings survive today in their original form. All we know of him is what his contemporaries said, which was often disparaging. Portions of Democritus's atomic theory, however, proved remarkably durable and stood without challenge for more than twenty centuries. Galileo and Sir Isaac Newton were believers in a universe made of circular points invisible to the eye, and Newton spent the latter part of his career in a fruitless quest to turn one element into another through alchemy. He never renounced his belief in the indivisibility of atoms, however, writing in his book *Opticks:* "It seems probable to me that God, in the beginning, form'd matter in solid, massy impenetrable particles . . . even so hard as never to wear or break into pieces, no ordinary power being able to divide what God Himself made one. . . ." The British chemist John Dalton, after conducting a series of experiments with gases in the eighteenth century, concluded not only that Newton was correct about a hard-balled universe but that atoms of a particular element were all exactly the same and that they were neither created nor destroyed.

The rapid discoveries about uranium at the beginning of the twentieth century, as well as the enthusiastic fictions of H. G. Wells, were challenging this ancient scientific orthodoxy. If uranium was tossing off particles that could be measured with an electrometer, then clearly a portion of matter existed that was even *smaller* than the atom itself. But what was it? And why was the uranium atom so eager to fling these specks?

The man who brought the world to the edge of these questions, and who did more than any one person to vivisect the atom, was Ernest Rutherford. The son of a poor New Zealand flax miller, Rutherford, at age twenty-four, won a scholarship to study physics at Cambridge and quickly amused his colleagues with his bluff antipodean manners. One colleague likened him to the keeper of a small general store in sheep country: "He sputtered a little as he talked, from time to time holding a match to a pipe which produced smoke and ash like a volcano." But the sodbuster exterior concealed a visionary mind, capable of drawing sublime inferences in the laboratory.

In 1906, Rutherford took a second look at Becquerel's rays, having previously categorized them into types. There was first the alpha ray, which could be stopped by skin or a piece of paper. There was then the beta ray—a free electron—which was negatively charged and could be blocked with an aluminum plate or a few sheets of paper. In order to understand the way the rays moved, Rutherford asked a graduate student to build a device that scattered particles from some of Marie Curie's radium through a narrow tube, after which they hit a screen coated with zinc sulfide. There they made a tiny spark that could be observed through a microscope. Further assistance was given by a student named Hans Geiger, who built a simple device out of a gas-filled tube and a thin metal wire that acted as an electrode. When a particle passed through this chamber, it set off an audible click. This became the prototype for the famous Geiger counter, the universal handheld radiation detector.

Rutherford put some gold foil at an angle to the zinc screen, which was out of the direct path of the radium.

What he witnessed was baffling. The alpha rays should have been passing through the foil. But a tiny percentage of them seemed to be bouncing off the gold foil and onto the zinc screen, in much the same way that a basketball will bank off a backboard and into the hoop.

"We found that many radiated particles are deflected at staggering angles—some recoil back along the same path they had come," said Rutherford. "And considering the enormous energy of the alpha particles, it is like firing a fifteen-inch shell at a piece of tissue paper and having it flung back at you."

He concluded that the alpha rays must have been bouncing off the hard core of the nucleus itself. Only one or two particles in a million were recoiling this way, suggesting that the nucleus was quite tiny and that only an extremely lucky shot could strike it—the equivalent of a hole-in-one golf shot. This suggested a much roomier atom than anyone had envisioned: a giant chamber of empty space with a positively charged center. If the atom were the size of a rugby field, its nucleus would be about the size of a strawberry seed.

"I was brought up to look at the atom as a nice hard fellow, red or gray in color, according to taste," Rutherford recalled later. That idea was now dead forever: There were now known to be interior gears that behaved in odd ways.

Rutherford had helped map the inner space of an atom, but he could not shake the idea that a ghost was hiding somewhere inside the structure. Protons and electrons announced their presence with a telltale electrical signal. But what if a fragment was lurking there that had no electrical aspect whatsoever? Invited to give a lecture to the Royal Society in London in 1920, Rutherford served up the scientific equivalent of a dead fish: an unsupported hunch.

There might be, he suggested, such a thing as an atomic particle "which has zero nuclear charge" and could therefore "move freely through matter." Such unrestricted movement through inner atomic space would have been out of the question for a proton or an electron, which would be bounced away from any surface because of its charge. But this phantom mote, if it existed, "should enter readily the structure of atoms, and may unite with the nucleus."

He was proposing a radical concept: that a particle released from one atom might slip past the shield of electrons and penetrate the nucleus of another, as a sperm fertilizes an egg. If this were the case, what would happen to the receiving atom in question? Would it change form? Would it explode?

One of Rutherford's assistants, a bespectacled twenty-nine-year-old named James Chadwick, resolved to find out if this unassuming particle existed. The riddle had been pushed along with some previous work on the part of Irène Curie, Marie and Pierre's daughter, who had been conducting experiments on the element beryllium alongside her husband, Frédéric Joliot. They had been firing alpha rays from polonium as "bullets" aimed at a sample of beryllium and witnessed the scattering of protons from the target—at three times the energy of the bombardment. The French couple made a key mistake, however, by concluding that this must be a form of gamma radiation, a type of electromagnetic wave.

Chadwick was known for his English reserve and milquetoast personality (he resembled, as one historian noted, a bit of a neutral particle himself), but he was irritated enough to shout, "I don't believe it!" when he read the results of the French experiment. He seemed to find the miscalculation almost offensive. Chadwick set out to replicate the beryllium experiment, except that he bombarded a host of other elements to show that protons were bumped out of them, too, and at a similarly rapid speed.

His conclusion was the reverse of the Joliot-Curies': The beryllium was not the source of the energy. Something lying near the heart of the alpha ray had to be causing the transfer of all that energy, and logic pointed to the lumpy ghost that Rutherford had predicted twelve years earlier. It was detectable only by the neighboring particles it caused to recoil. Rutherford himself later put it in vivid terms: It was like "an invisible man passing through Piccadilly Circus—his path can be traced only by the people he has pushed aside." Chadwick named this zero-charged particle the "neutron" and was consequently awarded the 1935 Nobel Prize in physics.

The basic map of the atom was nearly complete. And a few people around the world were beginning to grasp an ominous possibility.

"We have still far to go before we can pretend to understand the atom and the secret of matter," said the *New York Times* in a year-end roundup of scientific discoveries in 1932. "But we have gone far enough to think of an engine which will harness the energy released in atom building."

On September 11 of the following year, Ernest Rutherford gave the *Times* of London an interview in which he praised the uncloaking of the neutron as a giant leap forward. But he added that anyone who thought that useful power might be derived from neutron collisions was "talking moonshine."

A physicist named Leo Szilard happened to read this article while he was sitting in the lobby of a shabby London hotel. He became irritated, thinking Rutherford far too glib and blind to the destructive possibilities that the neutron suggested. Adolf Hitler had been appointed chancellor in Germany nine months before, and his ascendancy portended dark changes across the Continent. The next war, if it involved Germany, would likely turn on advances in technology, just as World War I had midwifed the tank, automatic rifles, and mustard gas. Szilard had been hounded out of his native Hungary by a rising tide of anti-Semitism, and he felt the world was becoming too dangerous to risk ignoring the military use of science.

Annoyed, Szilard left his hotel for a walk. He was standing at a stoplight in the Bloomsbury neighborhood, waiting to cross the street, when a bizarre and malevolent possibility occurred to him. He had dutifully studied Chadwick's results the year before, but perhaps more important, he had also just read H. G. Wells's *The World Set Free* and its fantastical account of a mineral that could be provoked into a "chain reaction" that would liberate the binding energy of heavy atoms all at once to create an inferno.

The neutron, thought Szilard as he stood in the London damp, was more than a piece of garbage. It would, in fact, be the perfect arrow to slice through the barrier of an atom's shell and directly engage the heart of the nucleus. This was not a new thought; Rutherford had predicted as much in his Royal Society lecture thirteen years before. But what if lobbing a neutron at the center of an atom resulted in the discharge of *two* neutrons that would, in turn, find their own nuclei to strike? The effect would be exponential, a riotous blossoming just as Wells had predicted: a recursive firing of component parts approaching the infinite halving of the flight of Zeno's arrow.

"As the light changed to green and I crossed the street, it . . . suddenly occurred to me that if we could find an element which is split by neutrons and which would emit two neutrons when it absorbs one neutron, such an element, if assembled in sufficiently large mass, could sustain a nuclear chain reaction," said Szilard, years later.

In another account of the same moment, he said he realized: "In certain circumstances it might be possible to set up a nuclear chain reaction, liberate energy on an industrial scale, and construct atomic bombs."

He was excited and horrified by this insight. The fictional had become suddenly possible, even likely.

"Knowing what this would mean, and I knew it because I had read H. G. Wells, I did not want this patent to become public," he said. "The only way to keep it from becoming public was to assign it to the government."

Szilard made his patent in the name of the British Admiralty, where it was received and promptly forgotten, despite his written warning that "information will leak out sooner or later. It is in the very nature of this invention that it cannot be kept secret for a very long time."

The discovery of the neutron revolutionized physics not only because it helped complete the diagram of the atom but also because it became an excellent tool for poking around the interior of different atoms, in much the same way that a Texas oil driller will sink a pipe into bedrock to see what lies below the surface.

At the University of Rome, the genial workaholic Enrico Fermi began bombarding the entire menu of the elements with neutrons to see what would happen. Lighter elements seemed impervious, but Fermi found that aluminum, once irradiated, transformed itself into an odd substance with a half-life of twelve minutes—an effect duplicated in heavier elements such as titanium, barium, and copper.

The strangest behavior of all was at the very top of the weight scale. When uranium was hit with neutrons, it ejected an electron and left behind a peculiar radioactive salad of unidentified elements with half-lives ranging from one to thirteen minutes. It would take time to sort out what had actually happened to Fermi's uranium, but it would eventually become clear that within the hash of metallic leftovers in his dish lay the secret of the atomic bomb.

Events began to move rapidly. The last half of the 1930s became a frenetic phase in physics as the study of uranium gripped laboratories on both sides of the Atlantic. The discoveries multiplied, like a chain reaction in itself.

"It was a period of patient work in the laboratory, of crucial experiments and daring action, of many false starts and many untenable conjectures," wrote the nuclear scientist J. Robert Oppenheimer, years after the fact. "It was a time of earnest correspondence and hurried conferences, of debate, criticism and brilliant mathematical improvisation. For those who participated it was a time of creation. There was terror as well as innovation in their new insight."

In Copenhagen, the great physicist Niels Bohr envisioned uranium's nucleus as "a wobbly droplet," an idea that helped explain why it was casting off pieces of itself. At the University of Chicago, Arthur J. Dempster discovered a rare version* of uranium with three fewer neutrons—thenceforth known as U-235—scattered through the rock like chips in a cookie. These atoms were more unstable than their neighbors, more likely to shatter if hit with a neutron. In Berlin, the research team of Otto Hahn and Lise Meitner aimed a stream of neutrons at a sample of uranium and found mysterious traces of middle-order elements such as barium and lanthanum inside the residue. In Paris, Frédéric Joliot-Curie found much the same thing.

And finally, at Christmastime in 1938, a major breakthrough arrived when a young Austrian professor named Otto Frisch, the son of a painter and a concert pianist, sat down for breakfast at a country inn with Lise Meitner, who happened to be his aunt.

Frisch had been working in a Hamburg laboratory before the Nazis took power in 1933. He had never cared much for politics, but the new anti-Semitic climate in Germany made it uncomfortable for him to stay, and he emigrated to London to take a teaching position. Frisch enjoyed whistling Bach fugues while at work in the laboratory and compulsively made pencil sketches of his colleagues during lectures; he later said the trick was to exaggerate their most noticeable features. He would have made a fine newspaper cartoonist had he not already been entranced with atomic physics.

---

* An element with a differing number of neutrons is known as an *isotope,* a Greek word meaning "in the same place."

Frisch was a shy man, and he found himself fumbling and stuttering when he was introduced briefly to Albert Einstein in the hallway of a university. But Frisch had a rare gift for a theoretical physicist—he was a superb classroom teacher who also spoke in plain language. His gift was not just of personality, it was one of dimensional visualization. Frisch knew how to conceive of invisible phenomena in vivid strokes, perhaps an extension of his knack for capturing the essence of a colleague's face in pencil by emphasizing a square jaw or bushy eyebrows. This talent was on full display when he joined his aunt for breakfast at the inn in Sweden where Meitner had been puzzling over a letter from Hahn in which he reported the presence of an uninvited mineral—barium—inside the wreckage of bombarded uranium. Hahn would later say that he had contemplated suicide when the true implications of this experiment became clear to him.

None of that guilt was present between Frisch and his aunt as they talked over breakfast and during a midmorning walk in the woods. Frisch suggested a novel idea. What if the nucleus was held together not so much by interior forces as by the electrical tension on the surface? Such a structure might be vulnerable to destruction when hit by a neutral particle, as the skin of a balloon is vulnerable to a needle. On a fallen log, the two stopped to rest, and Meitner pulled out a pencil and some paper she found in her purse. Frisch drew an oval that was grotesquely squashed in the middle to demonstrate the idea to his aunt—instead of funny faces, he was drawing funny atoms—and the two worked out some crude calculations.

This was the final untangling of the riddle of what was happening inside uranium when it admitted a neutron invader. Its center simply cracked into pieces, leaving behind radioactive fragments of its former self. The total mass was less than that of the uranium, meaning that part of it had escaped as pure energy. This explained the bizarre wreckage humming with fleeting half-lives that had so puzzled the Italian researchers in 1934. Frisch realized what Enrico Fermi could not have known at the time: The uranium atom had not transmuted, but actually had been *split* in his laboratory. This would seem to confirm what Albert Einstein had postulated in 1905: that even a tiny amount of matter could be converted into mammoth amounts of energy, tempered by the unchanging speed of light.

Frisch was at first doubtful of his own theory, reasoning that all the uranium deposits lying in the earth would have gone up in flames a long time ago if such a thing were possible. But now, he realized, after replicating Fermi's experiment, the reason they had not was because of the dilution of the unstable U-235 atoms. They occurred in quantities of about 0.7 percent inside natural uranium. All those neutron-spewing atoms were simply too far from one another to create a chain reaction. It would be like a drop of snake venom that loses its ability to kill when dissolved in gallons of water. But what if that venom could be distilled?

Frisch sent his results to the British journal *Nature*, which scheduled them for publication on February 11, 1939. The split halves of the uranium would be rushing apart at a speed of one-thirtieth the speed of light, he estimated: enough energy to make a grain of sand twitch from the popping of a single atom. This was a stupendous amount of force, perhaps as much as two hundred million electron volts, from such a small package. And if there was a large cluster of uranium atoms that started to pop? Two would create 4 would create 8 would create 64 would create 4,096. After eighty cycles of this, the number of exploding atoms would be a trillion trillion. It would all take place in less than one second.

Frisch borrowed a term from biology to describe the effect. When a single cell elongates and pulls apart into two, it is called fission. The word was appropriated to describe this new horizon of physical chemistry, invoking, as it did, a mysterious protosexual phase of life, majestic in its opacity. The spermlike neutron, unknown until recently, was the only thing that could pass through the armor of the electron shell and meet the heart of the nucleus. Frisch later wrote: "It was like possessing a magic arrow that would fly through the forest for miles until it found its mark."

In a later memo to the British government, he predicted that a brick of uranium no heavier than a gallon jug of milk would "produce a temperature comparable to that of the interior of the sun. The blast from such an explosion would destroy life in a wide area. The size of this area is difficult to estimate, but it would probably cover the center of a big city."

The fast-talking Leo Szilard meanwhile got himself hired on Fermi's team at Columbia University to work on the problem of chain reaction that had first haunted him at a London traffic signal.

He had been trying, fruitlessly, for the last six years to raise money for secret research and to keep the trick of atom splitting away from Nazi Germany. As part of his campaign, he wrote to the founder of the British General Electric Company, enclosing a copy of the prophetic opening chapter of Wells's *The World Set Free*.

"It is remarkable that Wells should have written those pages in 1914," he wrote, continuing with a lace of sarcasm. "Of course all this is moonshine, but I have reason to believe that in so far as the industrial applications of the present discoveries in physics are concerned, the forecast of the writers may prove to be more accurate than the forecasts of the scientists."

Szilard was impatient with the dithering, not just from the private sector but on the part of the governments in London and Washington, which had shown no interest in this fearsome quirk of nature that might either save the world or incinerate it. The future, as he saw it, was just as grim as H. G. Wells had foreseen. Scientific imagination was drawn immediately to warfare, and the primary use of the awesome power locked inside the atom would be for military ends. Conflicts could soon be waged—even possibly averted—with an otherwise unremarkable element at the top of the periodic table. Control of a peculiar glass dye from Bohemia would soon become a vital matter of national security.

In September 1938, Adolf Hitler had annexed the Sudetenland—a disputed border province of Czechoslovakia full of German speakers—sparking a diplomatic crisis that eventually led the British prime minister, Neville Chamberlain, to make the notorious statement that "peace for our time" had been secured. In the annexation, Hitler had unknowingly absorbed a jewel: the old mining town of St. Joachimsthal, one of the world's only known supplies of uranium. Should the Nazis also gain control of the diggings at Shinkolobwe, the United States and Britain would have none of the raw material necessary to construct an atomic weapon. The heaviest element in the periodic table was now widely believed to be the key to unleashing the thunderous force of binding energy. Newspapers were amplifying the news, even before Frisch's data could be published in *Nature*. Niels Bohr announced the results at a symposium in Washington, D.C., and the *New York Times* soon reported that "work on the newest 'fountain of atomic energy' is going furiously in many laboratories both here and in Europe. . . . It constitutes the biggest 'big game hunt' in modern physics." Luis Alvarez at the University of California at Berkeley read the news in a wire story reprinted in the *San Francisco Chronicle* while he was getting a haircut. "I got right out of that barber chair and ran as fast as I could to the Radiation Lab," he said. Uranium was suddenly in the international spotlight. Citing a January 30, 1939, press conference by the dean of the Columbia University physics department, the *Times* described the element as a "cannonball," capable of yielding "the greatest amount of atomic energy so far liberated by man on earth."

Thankfully, it was not that simple. Thousands of tons of uranium ore would have to be crushed, separated, and somehow enriched into a block of pure U-235 to develop the kind of jug-size bomb core necessary to flatten a city. This was a problem that transcended physics—it reached into geology, engineering, economics, and politics. There was no evidence as yet of a German atomic program, but Szilard thought it prudent not to waste time. He didn't know how to drive a car, so he enlisted a fellow Hungarian physicist, Eugene Wigner, to take him out to Nassau Point on Long Island, where the grand old man of science, Albert Einstein, had a summer cottage and was spending a few days sailing on Peconic Bay.

The pair got lost on a sandy lane and had to stop a small boy for directions to "Professor Einstein's house." Once inside the sitting room, teacups on their laps, the Hungarian visitors described the idea that had been spreading with viral speed among physicists since the publication of Frisch's article: that a slow neutron aimed at the center of uranium isotope 235 could trigger a splitting that would break the binding force and unleash two hundred million electron volts of electricity, as well as knock loose the fugitive neutrons that would instantly crack the neighboring atoms to create a massive, uncontrolled chain reaction.

Einstein's amiable reply, as recorded by Szilard: *"Daran habe ich gar nicht gedacht!"* ("I never thought of that!")

The resulting conversation in the summer cottage had less to do with physics than with politics and, in particular, with the existence of rock piles in Czechoslovakia and at Shinkolobwe. Szilard made a case that Einstein ought to alert his friend Elisabeth, queen dowager of Belgium, that the radioactive ore from her colony in the Congo ought to be transferred to the control of the United States.

Einstein was "very quick to see the implications and perfectly willing to do anything that needed to be done," recalled Szilard. "He was willing to assume responsibility for sounding the alarm, even though it was quite possible that the alarm might prove to be a false alarm. The one thing most scientists are really afraid of is to make fools of themselves. Einstein was free of such a fear and this above all else is what made his position unique on this occasion."

Einstein later allowed his signature to be affixed to the bottom of a measured letter written mostly by Szilard. It would ultimately not be addressed to the queen dowager of Belgium, but would be delivered by hand through an intermediary to President Franklin D. Roosevelt in the Oval Office on October 11, 1939. Though it does not mention Shinkolobwe by name, the threat in the Congo looms behind every sentence.

The letter, in full:

Sir:

Some recent work by E. Fermi and L. Szilard, which has been communicated to me in manuscript, leads me to expect that the element uranium may be turned into a new and important source of energy in the immediate future. Certain aspects of the situation which has arisen seem to call for watchfulness and, if necessary, quick action on the part of the administration. I believe therefore that it is my duty to bring to your attention the following facts and recommendations:

In the course of the last four months it has been made probable—through the work of Joliot in France as well as Fermi and Szilard in America—that it may become possible to set up a nuclear chain reaction in a large mass of uranium, by which vast amounts of power and large quantities of new radium like elements would be generated. Now it appears almost certain that this could be achieved in the immediate future.

This new phenomenon would also lead to the construction of bombs, and it is conceivable—though much less certain—that extremely powerful bombs of a new type may thus be constructed. A single bomb of this type, carried by boat and exploded in a port, might very well destroy the whole port together with some of the surrounding territory. However, such bombs might very well prove to be too heavy for transportation by air. The United States has only very poor ores of uranium in moderate quantities. There is some good ore in Canada and the former Czechoslovakia, while the most important source of uranium is Belgian Congo. In view of this situation you may think it desirable to have some permanent contact maintained between the administration and the group of physicists working on chain reactions in America. One possible way of achieving this might be for you to entrust with this task a person who has your confidence and who could perhaps serve in an unofficial capacity. His task might comprise the following:

a) To approach Government Departments, keep them informed of the further development, and put forward recommendations for Government action, giving particular attention to the problem of uranium ore for the United States;

b) To speed up the experimental work, which is at present being carried on within the limits of the budgets of University laboratories, by providing funds, if such funds be required, through his contacts with private persons who are willing to make a contribution for this cause, and perhaps also by obtaining the co-operation of industrial laboratories which have the necessary equipment.

I understand that Germany has actually stopped the sale of uranium from the Czechoslovakian mines, which she has taken over. That she should have taken such early action might perhaps be understood on the ground that the son of the German Under-Secretary of State, Von Weisacker, is attached to the Kaiser Wilhelm Institute in Berlin where some of the American work on uranium is now being repeated.

Yours very truly, Albert Einstein

Roosevelt handed the letter to his secretary Edwin "Pa" Wilson. "This needs action!" he said, and immediately authorized an ad hoc body called the Uranium Committee to examine the potential for building a weapon. But the Einstein letter did not work magic. There is no proof that Roosevelt even bothered to read it in full. The Uranium Committee was poorly funded (with an initial budget of $6,000) and led by career army men who were skeptical of the "magic bullet" the physicists were describing.

The British government, meanwhile—aided by the Bach-loving cartoonist Otto Frisch—was outpacing the Americans in both the imagination and the quality of the work. Frisch had been pondering the best way to remove the valuable U-235 from raw uranium and had arrived at the theoretical solution of mixing it with gaseous fluorine and forcing it through a tube with a heated rod in the center. If the tube walls were continuously cooled with water, the portion of the uranium with heavier isotopes would settle near the bottom while the lighter part could be harvested from the top. This method took advantage of the infinitesimal difference in weight between the quarry and its heavier relative, as when milk is separated from cream.

A simple idea, but extracting tiny amounts of U-235 in this fashion would be extremely time-consuming. A huge industrial facility would be needed to resolve the twenty-two pounds of U-235 judged necessary to achieve "critical mass"—that is, a lump of pure uranium so big that the neutrons would find more nuclei to smash than surface area to escape. This led to a joke among physicists. Why not just mail Adolf Hitler a dozen packages of uranium from different addresses? Each one would be brought to his desk for his personal inspection. When the last one arrived: Boom!

Critical mass, however, could never be achieved with raw uranium alone. And constructing a separation plant was a risky proposition in Britain, which was still in the bombsights of the Luftwaffe. Such a complex would consume enough electricity to light a city the size of Birmingham and cost up to $25 million, a sum the Crown could not afford to gamble on a theory, however promising. These recommendations were dutifully passed along to the United States, where they were systematically ignored.

"The minutes and reports had been sent to Lyman Briggs, who was the director of the Uranium Committee, and we were puzzled to receive virtually no comment," recalled the Australian physicist Mark Oliphant, who made multiple trips across the Atlantic in unheated bombers to shame "the cousins" into moving faster.

The complaints soon reached the ears of the White House, and control of what was called "the uranium question" was wrested from Briggs and transferred to the federal Office of Scientific Research and Development, an innocuous-sounding body that was responsible for adding new weapons, such as sonar, radar, and amphibious vehicles, to the American arsenal. Its director, Vannevar Bush, had been kept in the dark about Einstein's letter, but was quickly persuaded of the likelihood of a destructive energy release from a mass of enriched uranium.

And so, finally, the industrial and creative might of the United States began to awaken and apply itself to a crash program to build a uranium bomb. Bush was given authority to create a secret program code-named the Manhattan Engineer District, also known as the S-1 Project, and finally as simply the Manhattan Project. Its head would be General Leslie

Richard Groves, an arrogant but supremely competent administrative wizard from the U.S. Army Corps of Engineers who had recently overseen the construction of the Pentagon. That concrete star now behind him, he longed for a command position on the battlefield and was dismayed to learn of his assignment to steer what he viewed as a futuristic long shot.

At 250 pounds, with a truck-tire stomach, a wave of greasy hair, and a dead-fish handshake, Groves was a West Point man with a bookkeeper's thirst for minutiae and a mind that could graph a colossus from a few lines of statistics. He also had a total disregard for what people thought of him. His alkaline personality won him few friends (he was notorious for ordering colonels to pick up his dry cleaning), but it mattered little to him: He always favored prompt action over staff morale.

Groves's chief deputy, Kenneth D. Nichols, perhaps the closest thing he had to a friend, called him "the biggest son of a bitch I've ever met in my life, but also one of the most capable individuals. He had an ego second to none, he had tireless energy; he was a big man, a heavy man but he never seemed to tire. . . . I hated his guts and so did everybody else, but we had our form of understanding."

Groves had a mania for secrecy, and one of his first acts was ordering an information blackout, which extended to the popular media. One of the last stories to make it into print appeared in *Coronet* magazine in May 1942, under the headline URANIUM-235: CAN IT WIN THE WAR? The reporter, Murray Teigh Bloom, estimated there was barely enough of the "magic metal" in America to be piled on top of a dime, but that "there is every likelihood cheap, almost inexhaustible atomic power will be achieved in the lifetimes of most of us."

"I really got away with something," Bloom said from his retirement home in Connecticut. "It was an exciting period, and I followed a hunch. It was too big a thing to keep secret."

But with Groves in charge four months later, virtually all mentions of uranium disappeared from American technical journals. Editors were unwilling to disobey a Pentagon request in time of war, and Groves made a habit of paying rancorous personal visits to newspaper editors who ran with material he deemed compromising.

This was a gross overreaction. There was no real "secret" by this point. The physics of fission were public knowledge and well understood by scientists all over the world, as well as by the war departments of all the major combatants of World War II. "The bomb was latent in nature as a genome is latent in flesh," wrote the historian Richard Rhodes. "Any nation might learn to command its expression."

In Japan, the physicist Tokutaro Hagiwara lectured in 1941 on the possible development of a hydrogen bomb, using "super-explosive U-235" as the heating mechanism for the fusion of atoms. The Japanese army sent procurement officers to mines on the Korean peninsula to look for uranium. In Germany, the brilliant Werner Heisenberg envisioned using deuterium oxide—also called heavy water*—as a moderator to slow down the neutrons and create a more effective nuclear release. In France, a team of researchers led by Frédéric Joliot-Curie made a deal to secure fifty-five tons of uranium from Africa and discussed the possibility of testing an atomic weapon in the emptiness of the Sahara. In Russia, Igor Kurchatov had taken note of Frisch's article in *Nature* and told his government he feared that Germany or the United States would soon be collecting uranium, either for power or for weaponry. He later formed a committee to study ways to separate U-235; a senior deputy was moved to complain that younger Russian scientists "were so captivated by uranium projects that they forgot about the needs of the present day."

Yet all of these world powers were taxed and distracted by war and ultimately could not commit to an expensive theory, however promising it may have seemed. The embryonic nuclear programs in each of them suffered from lack of manpower, lack of money, lack of electrical power for isotopic separation, and, especially in the case of Germany, lack of support from the head of state. The führer was generally suspicious of technology and dismissed the idea of a uranium bomb as the "spawn of Jewish pseudo-science."

Building the bomb was now more of a bureaucratic matter than a scientific one. The job of cracking the subatomic code was finished. It now

---

* Winston Churchill, briefed on the theory, thought this term sounded "eerie, sinister."

came down to finding raw uranium in the ground, using brute industrial force to pull it apart, and then sculpting it in a precise globular shape for maximum fatal impact.

The physicist Ed Creutz expressed this recipe in stark terms during a meeting with a White House official. He made a cup with his hands about the size and shape of a baseball.

"All I need is a lump of uranium as big as this," he said. "But I need it now."

3

THE BARGAIN

Shinkolobwe was a perverse miracle, a globule of radioactivity that had burbled up from deep in the earth's crust five million years ago. Like most uranium deposits, it had seeped upward in carbonate solution and become trapped in the sinews of clays and granites. But its purity was more than two hundred times that of most uranium deposits. This would turn out to be a unique occurrence in the history of the planet, and now it stood as the best chance for the United States to gain a chokehold on world supply.

This prize had been found in the midst of the African bush in 1915 by the Belgian monopoly company Union Minière du Haut Katanga, which had inherited vast plains of mineral-rich territory from King Leopold. The company hired a swarm of inexperienced young men to explore the region, one of them an affable and unambitious Oxford graduate named Robert Rich Sharp, whose geological training had consisted of a single class. He took to the lifestyle immediately, and his native companions took to calling him Mlundavalu, which means "man who covers the country." On one of his treks, he happened to walk over the purest deposit of uranium anywhere in the world. Stories later circulated that Sharp had heard reports of African hunters smearing themselves with a colorful luminescent mud and had gone out to investigate. But Shinkolobwe's discovery was much more prosaic. While inspecting some properties of minor importance, Sharp climbed a short hill for a view.

"I was idly poking about on top when something yellow caught my eye," Sharp said later. He had seen uranium samples in a museum and now suspected the exotic mineral of creating the palette of colors on top of the hill.

Where uranium could be found, its daughter product, radium, would be sprinkled within. That spelled money. The boulders scattered at the base could have been worth as much as $2 million each. Sharp found a zinc plate, used a penknife to scratch out the word RADIUM in letters six inches high, and mounted the sign atop the hill. A sample of the yellow rock was sent in for assay, and when it indeed proved to be uranium, the claim was given the name of the nearby village: Shinkolobwe, "the fruit that scalds."

A black workforce was recruited and contained inside a sealed compound, near a manager's village of brick houses and streetlights. The mine opened for business in 1922 and began to flood the market with medicinal radium, putting the competition at St. Joachimsthal nearly out of business. All the profits flowed upward, first to Union Minière and then to its mammoth holding company, the Société Générale, which had inherited most of King Leopold's plantation and ran it with a lighter touch, albeit with the same forced-labor policies.

"The Congo can best be understood as the private preserve and reservation of S.G. [Société Générale]," noted one American intelligence brief. "For all major practical considerations, S.G. *is* the Congo." Locals in Katanga often spoke of the company in terms interchangeable with the colonial government, and its influence over the financial houses of Europe was said to have equaled that of the Rothschild family or the J. P. Morgan banking empire.

The man who had oversight of the Congo's new treasure was Edgar Sengier, the portly and dapper director of Union Minière. He had started his career as a tramway engineer in China and had risen to the top of the Belgian mining giant. Sengier belonged to the best gentlemen's clubs in Britain and France and enjoyed excellent wine wherever he went. His suits were bespoke and always sparkling clean. His skin was china pale, he walked with a slight mince, and his silver mustache was always trimmed sharp; it was his custom to send a bouquet of pink carnations to the wives of men with whom he dined. Sengier was sometimes described (with only partial exaggeration) as one of the most powerful men in the world. He had direct control of 7 percent of the world's copper and almost all of its cobalt. And every aspect of this empire was subject to his micromanagement. "Never allow a lawyer to draw up a contract," he once confided to an acquaintance. "Always write it yourself. Then you will know exactly what it means."

As Europe began sliding toward war in the late 1930s, the market for radium began to suffer, and Sengier closed down the Shinkolobwe mine. He neglected to have it pumped, and the pit flooded with dirty water. A visitor described it as "a gray ulcer." Piles of surplus ore, which had been painstakingly sorted by Congolese hands, sat in a nearby warehouse. The drills and carts were transferred to nearby copper mines; the ore muckers were all fired and sent home.

Sengier would learn that his Congo property could turn out to be interesting after all. During a trip to London in 1939, he was introduced, through a friend, to Sir Henry Tizard, the director of the Imperial College of Science and Technology, who had been briefed about uranium's propensity to undergo chain reaction. He asked Sengier, casually, if he would be willing to give the British government an option on his inventory, which then amounted to nearly six thousand pounds. Tizard made only passing reference to the encounter in his diary, and the price he named is lost to history, but Sengier must not have thought much of it because he refused the offer and did not attempt to haggle. But Tizard made a memorable remark at the end of their meeting, one that Sengier never forgot. "Be careful, and never forget that you have something which may mean a catastrophe to your country and to mine if this material was to fall into the hands of a possible enemy."

This was a strange comment: If the British really believed this to be so, why were they not offering a better price?

Sengier's suspicions were reinforced only a few days later when a delegation from France, led by Joliot-Curie, came to see him with an ambitious idea to test a uranium bomb in the Sahara. They offered generous terms, which he accepted: Union Minière would receive half the royalties on any patents resulting from their experiments. This deal was quashed after the Nazis invaded Belgium in May 1940. Heaps of yellow uranium ore on the docks became the immediate property of Adolf Hitler.

But Hitler missed acquiring the largest part of Shinkolobwe's inventory, which in a lucky accident for the United States had already been hidden inside a warehouse in New York City.

Edgar Sengier hated the Nazis and had guessed they would probably invade Belgium on their way into France. He also guessed, correctly, that war would be excellent business and that he could best conduct his trade in the United States, which would soon become the world's biggest user of cobalt, a vital metal for the assembly of aircraft engines. Sengier rented a permanent suite for himself and his wife at the Ambassador Hotel in New York and set up an office-in-exile in the Cunard Building at 25 Broadway for a front company called African Metals Corporation. His native Belgium might now have been decorated with swastikas, but his company's grid of mines, mills, and railways in the Congo was still operational and eager to do business with the war machine stirring itself to life.

Sengier did not forget the uranium. He arranged the barreling of the remaining inventory at Shinkolobwe—about 1,250 tons—and had it railed to the port at Lobito. This was done without notice or fanfare to throw off any Nazi informants. The barrels were loaded onto two separate freighters and taken across the Atlantic to Staten Island, New York, where they were stored in a three-story warehouse on the site of a vegetable oil plant run by Archer Daniels Midland, near the southern footing of the Bayonne Bridge. It was the only place Sengier's deputies could find on short notice. Each barrel was stamped with the plain legend URANIUM ORE—PRODUCT OF BELGIAN CONGO.

The lethal mineral would sit in this obscure corner of the harbor, unwanted, for more than two years. Sengier made halfhearted attempts to sell the lot of it to the U.S. government, which was not interested. Days after the bombing of Pearl Harbor, he met with a State Department official about the urgent need for more cobalt. Sengier told his visitor that he really ought to have been concerned with Union Minière's leftover uranium, and perhaps lock it up with the nation's gold supply inside Fort Knox. But the official had not been briefed on the element's destructive capabilities, and he could only equivocate. Two follow-up letters sent by Sengier were also brushed off. But with the ursine Leslie Groves now in charge of the Manhattan Project, a different kind of meeting finally took place.

With an enormous budget at his disposal and the backing of the White House, Groves would soon be on his way to making deals with some of the largest chemical and engineering corporations in America: Bechtel, DuPont, Raytheon, Eastman Kodak, and Union Carbide would all be hired to erect the continental apparatus needed to produce the atomic bomb. But raw uranium was the first concern. The only domestic supply was inside old slag heaps in the Colorado mountains. Union Minière had effectively killed the American radium business twenty years before, at the same time it destroyed the prominence of St. Joachimsthal.

Groves sent his chief deputy, Lieutenant Colonel Kenneth D. Nichols (the one who privately considered his boss a "son of a bitch"), to Manhattan on September 18, 1942, to buy whatever uranium he could from the Belgian company. An elegant man in his sixties was there to greet him.

"He had a somewhat pallid face and his light hair was thinning," recalled Nichols. "He was immaculately dressed and he spoke excellent English in rather curt sentences."

After inspecting the colonel's military ID, Sengier asked with an acid tone: "Are you a contracting officer? Too many people have been around here about this uranium, and they just want to talk. Do you have any authority to buy?"

"Yes, I have more authority, I'm sure, than you have uranium to sell," said Nichols.

"Will the uranium be used for military purposes?" Sengier demanded, and Nichols hesitated, knowing he could not discuss the secret project.

"You don't need to tell me how you'll use it," Sengier said. "I think I know. When do you need it?"

"If it wasn't impossible, I'd say tomorrow."

"It's not impossible. You can have immediately one thousand tons of uranium ore." The Belgian then told him of the unguarded stockpile sitting barely eight miles away, inside a vegetable oil plant.

"I have been waiting for your visit," he added, and took out a yellow legal pad and a pencil.

Nichols left the office thirty minutes later, carrying a sheet of scratch paper that formed the beginning of a covenant between Union Minière and the United States that would last for the next eighteen years. The barrels of uranium were immediately taken to a military depot in New Jersey. Additional shipments from Shinkolobwe were ordered, an average of four hundred tons of uranium oxide per month loaded onto fast freighters that could outrun German U-boats. Only one of them would ever be lost to torpedoing.

Payouts to Union Minière were made through a dummy account at Bankers Trust Company, which at one point contained $37.5 million. "There was to be a minimum of correspondence on the subject and the auditors were directed to accept Sengier's statements without explanation," noted Groves. At $1.04 per pound, the price was slightly inflated, but Groves had a bottomless budget, and uranium now appeared to be the possible savior of the Allied cause and the energy source of the future. Groves said later in his memoirs, "Its value had never been determined in the open market, and now there was only one purchaser and one seller."

With the New York barrels safely locked down, the United States set about denying uranium to the rest of the world. It was thought at the time that uranium was a geologically rare element, found only in select locations. He who controlled the uranium deposits, therefore, ought to be able to control the world after the war was over. This was the philosophy behind a clandestine survey of global uranium reserves, conducted for the Manhattan Project by Major Paul L. Guarin, a Texan who had worked as a geologist for Shell Oil before the war and possessed a swashbuckling temperament that suited Groves, who remarked, "I did not want anyone who would always insist on 100 percent proof before making a move." Under the code name "Murray Hill Area," after a neighborhood in Manhattan, Guarin hired consultants from Union Carbide to comb about sixty-seven thousand geological volumes, assay soil samples from twenty friendly or neutral countries, and study new methods for identifying uranium ore deposits. Nations were then cataloged according to their uranium-bearing potential. The only one that rated "excellent" was the Belgian Congo. Listed as "good" were Canada, the United States, and Sweden;

and judged "fair" were Czechoslovakia, Portugal, and South Africa. The USSR and Bulgaria were marked as "unknown."

"All other countries appear to have very poor production possibilities," concluded the final report. This was based entirely on Leslie Groves's working assumption that just a few places in the earth's crust—such as Colorado and Shinkolobwe—had been endowed with the volatile element. He believed that even the massive territory of the USSR concealed no appreciable uranium and, as a result, that it would likely take the Soviets at least twenty years to build their own atomic bomb. Cornering the world's uranium, therefore, meant that the United States should be able to preserve an atomic monopoly for decades to come.

This was a bad miscalculation. Uranium turned out to be more common than tin, and nearly five hundred times more abundant than gold. At least a hundred billion tons of reserves are now known to exist, including substantial holdings in Russia. Richard Rhodes summarized Groves's quest thusly: "He might as well have tried to hoard the sea."

There was not nearly enough ore in the New York barrels to make a bomb, and so Sengier had to be persuaded to reopen the Shinkolobwe mine, which was full of dirty water. The Belgian showed himself to be a tough negotiator; humorous but brittle. "Well, General," he asked Groves at the beginning of their meeting, "are we going to play poker or are you going to show your cards?"

Sengier refused the initial offers, but was persuaded to change his mind in 1943 after he was offered an exclusive buying contract and the free construction assistance of the U.S. Army Corps of Engineers, which would drain the water from the pit. He thereby managed to squeeze out a major financial gift. The United States would pour $13 million into retooling Shinkolobwe—in effect, subsidizing a global monopoly on uranium for the Belgian owners.

A U.S. Army private named Joe Volpe was sent out to inspect the property, and he found the mine office in Elisabethville full of uranium rocks. They were samples designed to impress visitors with their colorful staining.

"Don't you boys know that this stuff will make you sterile?" he asked, only half joking.

The Belgian managers replied—somewhat defensively—that they had already fathered several children. But when Volpe returned to the office on another visit, he saw that the samples had disappeared.

A unit of the U.S. Office of Strategic Services was sent into the Congo to watch the site for any signs of Nazi sabotage. A young diplomatic officer named Robert Laxalt recalled meeting the head of the unit in Léopoldville in 1944, a man with a "sphinx face" and "the most piercing eyes I have ever encountered." The spy imprudently revealed his mission one night: "There's something in that mine that both the United States and Germany want more than anything else in the world. The Shinkolobwe mine contains a mineral called 'uranium.' The Congo has the only producing mine."

British agents were also lurking inside Czechoslovakia during the war, monitoring St. Joachimsthal for any signs of large-scale digging. This was thought to be a sure sign that Hitler, too, had made progress on an atomic bomb. But that trip wire was never activated. "Tailings piles from each mine were microscopically measured from one reconnaissance to the next," wrote Groves. "There were no signs of extraordinary activity."

The Germans had not made much of the uranium already in their possession. At a laboratory in Leipzig in 1942, Robert Dopel and Werner Heisenberg managed to construct a crude spherical fission device out of uranium and heavy water. But it started to leak, and when the physicists opened the outer shell for inspection, the uranium reacted with the air, caught on fire, and then burst in a harmless nonnuclear fizzle, spraying the whole lab with a mess of burning uranium that set the building ablaze. The Leipzig fire brigade offered the pair congratulations for the achievement in "atomic fission," leaving the scientists in despair. This was one of the only known uses of St. Joachimsthal uranium during World War II and high-water mark of the Nazi nuclear effort.

Shinkolobwe was a much bigger prize. Parts of its ore body demonstrated a freakishly high grade of 63 percent uranium. Moreover, it could be operated in the secrecy and obscurity of the African heartland, with a ready workforce close at hand and within the borders of a friendly colonial power. It is doubtful the Manhattan Project ever would have developed a bomb without Shinkolobwe. Even its garbage was a treasure. The geologist Phillip Merritt was sent out to the mine for a look in 1943, and he found ore in the waste piles that registered up to 20 percent pure uranium, far surpassing anything else that could be mined in the world. High-grade ore began to flow into the United States, some of it packed in burlap bags left over from a South American tin mine, each bag stamped with the legend PRODUCT OF BOLIVIA.

Once the uranium was inside America, the fissile component had to be separated from the more stable part of the ore, and this would take a gigantic amount of electricity. One possible method had been suggested by John Dunning and Eugene Booth at Columbia University, and it built upon the concept proposed by Otto Frisch in Britain: Mix the uranium with hexafluoride gas and pump it against a porous surface (a screen made of millions of tiny openings) that would capture the lightest part. If repeated thousands of times, this would create a "cascade" effect that would eventually yield enough U-235 to shape into a bomb.

But uranium hexafluoride was incredibly corrosive, a gassy version of battery acid, and it would take thousands of high-quality separation tanks all working in succession to push it through the cascades and eke out even a few pounds of the necessary material. Such a pharaonic project would require an isolated patch of countryside that also happened to be near a source of cold water and a large electrical facility.

The Manhattan Project condemned just such an area of fifty-nine thousand acres along Tennessee's Black Oak Ridge near the Clinch River that happened to be nearby a brand-new TVA power plant and far away from prying eyes. Two small towns were evacuated and demolished; the region, now called the Clinton Engineer Works, was sealed off; and construction began on the gaseous diffusion plant. Dubbed K-25, the plant would employ twelve thousand people, who were housed in a muddy settlement called Oak Ridge, which was itself nicknamed "Dogpatch" after the hayseed town portrayed in the newspaper comic strip *Li'l Abner*.

The diffusion plant was shaped like a large *U*. Each of its legs extended a half mile; technicians found it convenient to use bicycles to travel from one end to the other. It was, at the time, the largest building anywhere in the world. Another isotopic separation plant on the site, a racetrack-shape series of electromagnetic separators known as calutrons, went online early in 1945. It was located in a valley seven miles from the other plant, under the logic that if one should explode, the other would still be functional. The finished uranium-235 was stored in a hollowed-out bluff near a white farmhouse, the grain silo of which was actually a machine-gun nest.

Most of the employees were never told that their jobs were connected to weapons. Those who knew of the presence of uranium were never supposed to call the stuff by its real name; official nonsense words—*tuballoy* and *yttrium*—were coined instead. Even senior managers had little idea of what was happening. This was a Groves hallmark: He was a compulsive hoarder of information, keeping most of his command chain divided into separate units forbidden to communicate with one another. "Every man should know everything he needs to know to do his job, and nothing else" was one of his maxims, and the culture of secrecy infected every corner of the Manhattan Project. The officer in charge of contracts with Union Minière became frustrated with the "cryptic conversations" he was forced to have with the Belgians and complained about the baffling conversations he was forced to interpret.

Yet another facility was located on the grounds at Oak Ridge: a tomb-like structure of concrete nicknamed the "Black Barn" and officially called X-10. This was America's first permanent nuclear reactor, which was designed to synthesize a newly discovered element called plutonium. A close cousin of uranium, and even more fissile, it had been isolated in 1941 by a team of researchers at the University of California led by the physics professor Glenn Seaborg, who had successfully bombarded uranium with neutrons to produce the first "transuranic" element; the same element that Enrico Fermi had tried to discover in Italy. Plutonium's name was foreordained. It came from the planet Pluto, which had first been spotted from an Arizona observatory only eleven years prior. This was a nod to the long-ago German pharmacist Martin Klaproth, who had christened uranium for a distant member of the solar system.

Plutonium does not occur in nature, except as a freak occurrence and in minute quantities. It is formed when uranium is bombarded with neutrons, thus creating an element that oxidizes in eerie pinkish-orange colors and has virtually no use except widespread destruction. It emits so many alpha rays that even a small chunk of it is warm to the touch. The Manhattan Project scientists were unsure if it would be as deadly as uranium in a bomb, but Enrico Fermi had already demonstrated a dependable way to manufacture it in bulk.

Fermi had built and tested the world's first nuclear reactor on an old squash court underneath Stagg Field at the University of Chicago in December 1942. The "pile," as he called it, was a simple structure of uranium slugs encased with graphite bricks to slow down the neutrons. Size was the crucial factor: It had to be large enough to achieve critical mass but small enough to avoid flooding the South Side of Chicago with a wave of radiation. This required some ingenuity. Fermi's team had designed rods made of cadmium (an element that absorbs neutrons) that could be lowered into the bricks to calm the eruption of the uranium atoms. The chain reaction could therefore be ignited and snuffed at will. A nuclear fission, once thought to be physical anarchy, turned out to be as easy to command as a propane flame on a barbecue grill.

This would become the basic model for all nuclear power plants, and the X-10 at Tennessee was designed to replicate the feat. Except that the creation of plutonium, not energy, would be the true purpose.

Groves decided to make an expensive bet on the new element and ordered the purchase of a sere and depopulated region of Washington State near the Columbia River. "Most of the area was sagebrush, suitable only for driving sheep to and from summer pasture in the mountains," said Groves. A few families had lived there since their ancestors had come to the Northwest in horse-drawn wagons, and more than one kept a hearth fire that had first been lit in the prior century and kept aflame for sentimental reasons. When the army bought their homes, it was forced to scoop up and transport a few cheerily burning fires as well.

The U.S. government would, in secret, turn this land into the Hanford Site: a reservation for the manufacture of plutonium that would soon become the most polluted piece of real estate on earth. It was half the size of Rhode Island and featured 3 nuclear reactors, 540 buildings, a wartime budget of $358 million, and more employees than it had taken to dig the Panama Canal four decades earlier. A rock chipped out of the ground by farmers living in near Stone Age conditions was fed into the most advanced industrial complex ever constructed, an endeavor the budget and employment figures of which were, according to the historian Richard Rhodes, on rough par with the entire automobile industry. Its only task was to process what little uranium the United States had managed to secure from three sources: Edgar Sengier's Shinkolobwe, the tailings from old radium mines in Colorado, and a mine on the shore of Canada's Great Bear Lake that had been discovered by a skilled woodsman named Gilbert LaBine who had set out to the Arctic in 1929 to get rich off the craze for radium. Winter temperatures there were in the minus-fifties and an American miner named Fred Chester Bond marveled at how the mine walls were covered in ice. Shining a lantern inside made everything sparkle into seeming infinity; like "a children's fairyland," he said. This Arctic mill was reopened in a hurry in 1942 upon the demand of Leslie Groves to the Canadian government and would go on to provide nearly one-third of the radioactive bullion for the Manhattan Project, which had been combing the world for every available pound.

By the spring of 1945, Japan's surrender was becoming increasingly certain, and it remained doubtful that the United States would be able to produce a usable atomic bomb by the end of the war. There were only thirty-three pounds of enriched uranium available, not nearly enough to achieve the target of 2.8 critical masses that a device would require. Leslie Groves pressured his DuPont contractors at Hanford to boost the plutonium output; his subordinates termed this "the super-acceleration program." The majority of America's stock of natural uranium had already been channeled into the reactors at Hanford, and this judgment would come as a costly embarrassment if the fuel should fail to materialize by the end of the war. An acquaintance of Groves's joked that Groves ought to buy a house near Capitol Hill—so it would be an easy walk to Congress to answer for all his promiscuous spending.

Then came a surprise.

A little over a month before Adolf Hitler ate a last meal of spaghetti and shot himself in the temple in his bunker near Wilhelmstrasse in 1945, a submarine named *Unterseeboot-234* sailed out of the harbor in Kiel, Germany, with a crew of sixty and some very curious cargo.

The vessel was one of the biggest submarines in the German fleet, nearly three times as large as an average U-boat. It had left Germany in a twilight atmosphere: The Red Army was closing in on Berlin, and the Americans under General Dwight D. Eisenhower had already crossed the Moselle River into the heart of the disintegrating Reich. The submarine's captain was Johann Fehler, a lieutenant who had never before seen undersea combat. He was the best the German military could find in their ravaged ranks. Allied depth charges or torpedoes had already killed nearly three out of every four submariners. "It was clear to all of us that this war was lost, and nobody wanted to be a part of this mission," said one of the crew.

Its orders were to ship to Japan a load of sensitive military equipment, including proximity fuses, blueprints for the V-2 rocket, chemical weapons, and two complete jet fighters, which had been dismantled and their parts wrapped for the voyage. And tucked away in the box keel was the primary cargo: a series of ten wooden boxes stamped ST 1270/1-10, JAPANESE ARMY, each holding metal cylinders lined with gold foil and containing powdered uranium oxide.

There were 1,235 pounds of it in all, the remnants of Germany's halfhearted attempt to build its own atomic bomb. The uranium had possibly come from the seizure of Union Minière's yellow pyramids on the docks in Belgium, but more likely it was from the tailing dumps at St. Joachimsthal. It was now being shipped out in this eleventh-hour attempt to pass usable war matériel to Japan.

The submarine was also carrying fourteen passengers, mostly high-ranking Nazi military officers who, in the face of certain defeat at home, had been ordered to aid Japan in her foundering struggle in the Pacific. Among them were the Luftwaffe general Ulrich Kessler, an expert in air defense, and Dr. Heinz Schlicke, a rocket scientist who was supposed to help Tokyo manufacture its own V-2 rockets. Two high-ranking members of the Japanese navy were also on board, Genzo Shoji and Hideo Tomonaga. It was not clear which of the passengers, if any, had been assigned to safeguard the uranium.

Captain Fehler initially set a course that would take the sub to Southeast Asia by going around the tip of Africa and through the Melaka Straits. He avoided enemy destroyers by descending to a level of nine hundred feet and sneaking unscathed into the North Atlantic, even though British cryptographers were monitoring the submarine's transmissions. But on May 4, the entire crew heard a special shortwave broadcast from Admiral Karl Dönitz, who had been appointed president of the shattered Reich after Hitler's suicide. His only job was to surrender the remnants of the German war machine to the Allies, and he started with his own navy.

"My U-boat men," he said. "Six years of U-boat warfare lie behind us. You have fought like lions. A crushing superiority has compressed us into a narrow area. The continuation of the struggle is impossible from the bases that remain. U-boat men, unbroken in your war-like courage, you are laying down your arms after a heroic fight which knows no equal."

Several members of the U-boat crew argued for ignoring the order and hiding out in Argentina or on an island in the South Pacific, using their military cargo as something to trade for food and new clothing. Captain Fehler saw things differently. The sub would be considered a pirate vessel if he did not surrender it at the first opportunity. But to whom? Russia was out of the question. The Red Army was known to be vengeful for all the misery it had suffered at Hitler's hands; its cruelty to surrendering troops was already legendary. Britain was also rejected, due to the ferocious blitz bombings it had suffered earlier in the war and the suspicion that the British would turn submarine POWs over to the French. The United States was seen as the best choice. It had no long history of militarism, and its soil had been unsullied by German bombing.

Fehler changed his course and started heading west across the North Atlantic. This decision did not sit well with the two Japanese officers, for whom surrender was considered a great dishonor. The war may have been over for the Germans, but it was not for them. Emperor Hirohito had not given the order to quit, and the homeland was still under attack. Shoji and Tomonaga each took an overdose of sleeping pills and died quietly in their bunks, family photographs at their side. Their German companions did not interfere with their hara-kiri ("It was their right," one reasoned) and buried them at sea.

On May 12, after radioing a set of false positions to the Canadians, Fehler's submarine officially surrendered to the USS *Sutton* and was boarded by a squad of soldiers, many of whom were amazed to see a German U-boat up close. The vessel, its crew, and all of its contents arrived at the Portsmouth Naval Yard on May 19. Newsreel photographers filmed the crew being escorted off the giant submarine, with General Kessler getting most of the attention on account of his iconic Prussian bearing, complete with monocle, Iron Cross, and long leather coat.

Fehler protested the media circus around the gangplank, and a U.S. Navy captain showboated in front of the newsreel cameras by yelling at the "Nazi gangsters" to "get off my ship!" The crew was taken away to prison cells for interrogation. Almost immediately, the submarine watch officer, Karl Pfaff, revealed what he knew about the uranium, which he believed to be "highly radioactive."

After hearing of this, the Office of Naval Intelligence radioed an order to Portsmouth on May 27: "Uranium oxide loaded in gold-lined cylinders and as long as cylinders not opened can be handled like crude TNT. These containers should not be opened as substance will become sensitive and dangerous." The shipment was apparently taken off the ship without incident. A translated version of the official manifest shows "560.0 [kilograms] uranium oxide, Jap Army" among the contents. The containers were shipped to a navy lab at Indian Head, Maryland, for testing, and Pfaff later acknowledged he was on hand to help American crews safely open the containers with blowtorches.

At that point, the uranium disappeared from the record. A curtain of secrecy descended, and the navy issued no further documents about the cargo. The transcript of the radio call to Portsmouth marks the end of the paper trail.

What happened to Germany's twilight shipment of uranium? There is no definitive consensus among historians of World War II who are familiar with the incident, but one logical outcome seems to be that it was delivered to the Manhattan Project for use in weapons development. There would have been no other use for such cargo, particularly at a time when America's atomic effort was scrambling for every last crumb of the element it could find. Physicists had discovered just two months earlier that they would need even more of it than forecast—up to 110 pounds—to achieve the necessary yield in the first blast. General Leslie

Groves would have certainly been notified by U.S. Navy officials once the submarine cargo's true nature was discovered. His voracious appetite for uranium would have made storage or destruction an unlikely possibility. The unexplained information blackout after May 29 also bears the marks of his characteristic insistence upon secrecy at every level.

A possibility thus hangs over the uranium cargo: It may have become part of the bomb detonated over Hiroshima on August 6, 1945, and thus Japan would have eventually received its uranium, albeit three months late and in a different form.

In 1995, the chief of security for the Manhattan Project came forward and acknowledged that he had indeed ordered the uranium to be delivered for use in an atomic weapon.

John D. Lansdale was a native of California and a graduate of Harvard Law School. He had been working as an attorney in Cleveland at the outbreak of the war, when he joined the U.S. Army as an officer. Groves eventually chose him to be the head of security and intelligence for the Manhattan Project, and Lansdale spent much of the war trying to determine the extent of the German bomb program. He also gave security clearances for top personnel, including Paul Tibbets, the Army Air Corps pilot chosen to fly the B-29 that would drop the first bomb on Japan. When the Nazi submarine was escorted into New Hampshire, Lansdale said he became the man in charge of disposing of its unexpected gift of uranium oxide. There was no doubt in his mind as to what happened to it.

"It went to the Manhattan District," he told William J. Broad of the *New York Times*. "It certainly went into the Manhattan District supply of uranium."

Lansdale could provide no documentation to support his claim, but he elaborated further in a videotaped interview given to a documentary filmmaker shortly before his death in 2003. He was in frail health at the time, and barely audible at points, but he was unequivocal about what had happened to the cargo.

"When I heard about the uranium aboard the German submarine, I got very excited because I knew we needed it all," he said. "I made arrangements with my staff to retrieve and test the material. I sent trucks to Portsmouth to unload the uranium and then I sent it to Washington. After the uranium was inspected in Washington, it was sent to Clinton. . . . The submarine was a godsend because it came at the right place at the right time."

By "Washington," Lansdale was likely referring to the U.S. Navy Ordnance Investigation Laboratory at Indian Head, Maryland, where Pfaff had helped unseal the containers. "Clinton" is a clear reference to the Clinton Engineer Works at Oak Ridge, Tennessee. If the uranium had been shipped by mid-June, there would have been ample time for technicians to process it into components that made up the world's first atomic bomb.

But a brief note in Leslie Groves's administrative diary on August 13, 1945, seems to suggest that while the German uranium was indeed sent to the complex at Oak Ridge, it never made it into the Hiroshima bomb, which had been exploded seven days earlier. The entry for 10:33 a.m., typed by Groves's secretary, summarized the contents of a telephone call from a top navy officer: "Admiral Edwards just called to ascertain if the material we got from the German submarines was of any use to the program. General advised it wasn't of any help as yet, but that it would be utilized." As a compulsive micromanager, Groves certainly would have had knowledge of such a detail, and he would have had no known motivation to hide the truth from a U.S. Navy admiral during a secure phone call.

Though he had no knowledge of the ultimate fate of the Nazi uranium, the Manhattan Project physicist Hans Bethe believed that nobody would have objected if it had been put into the production stream for the Hiroshima weapon. The race to build America's bomb was closely linked with the hunger for its basic material.

"We wouldn't care where the uranium had come from," he said. "We wanted all we could get."

The scientific laboratory of the Manhattan Project was the campus of a boys' school in the mountains of northern New Mexico. It was in the midst of the favorite horseback-riding country of the head of the scientific team, thirty-eight-year-old J. Robert Oppenheimer, who had grown up in a privileged household on New York City's Riverside Drive. He was a self-described "unctuous, repulsively *good* little boy" who went on to teach physics at the University of California at Berkeley. "Oppie," as he was known, was lanky and blue-eyed, with a taste for martinis, cigarettes, and spicy food. He could be callous to men he considered lesser than he, but he harbored delicate passions: Renaissance poetry, Eastern mystical religions, and the yawning beauty of the deserts. His first words upon hearing of the discovery of uranium fission in January 1939 had been "That's impossible!" But within the afternoon, he had grasped the possibilities and agreed that such an event would have enormous capabilities. Within the week, a crude sketch of a bomb was on his blackboard.

He and General Groves recognized something in each other—a supreme competence—but clashed on some of the details of exactly how the best scientific minds in the world would be assembled in a remote location and set to work on the mechanics of the atomic bomb. Groves was not shy about sharing his opinion of the physicists under his command. He thought they were mostly "crackpots" and "prima donnas" (and he harbored an intense dislike of the unruly Leo Szilard), but he agreed to purchase the campus of the Los Alamos Ranch School on top of a piney mesa to keep his scientific chief happy. "I am the impresario of a two-billion-dollar opera with thousands of temperamental stars," he liked to brag. Oppenheimer managed to talk him out of making all the scientists wear military uniforms, as well as a daily bugle drill in which everyone—Nobel laureates and all—would have been roused at dawn to scan the skies for enemy parachutists.

A small army-built city with laboratories, a mess hall, a movie theater, and apartments rose on the mesa. Some local Pueblo Indians were hired for construction tasks, inspiring curiosity among some of the foreign-born contingent of academics who staffed Los Alamos. "There they were, the oldest peoples of America, conservative, unchanged, barely touched by our industrial civilization, working on a project with an object so radical that it would be hailed as initiating a new age," recalled the wife of one scientist.

All the mail was sent to a catchall address—PO Box 1663—and residents were strictly cautioned against telling their relatives exactly where the army had relocated them. "A whole social world existed in nowhere in which people were married and babies were born nowhere," noted one observer. The patriotic sensibilities of the times meant that the locals asked few questions about the explosions echoing off the canyon walls. They were really tests of the conventional dynamite needed to trigger atomic fission in a plutonium bomb, but the general assumption was that Los Alamos must be a secret munitions plant or a factory for poison gas. When Thomas Raper, a vacationing reporter from the *Cleveland Press*, showed up at the gates to do a story, he was firmly turned away by the guards. But it did not stop him from assembling a pastiche of local rumor under the headline FORBIDDEN CITY once he returned to Ohio. He wrote, "The Mr. Big of the city is a college professor, Dr. J. Robert Oppenheimer, called 'the Second Einstein' by the newspapers of the west coast." Groves reacted with his usual asperity and made inquiries about having the reporter drafted for military service in the Pacific. He gave up when he learned that Raper was almost a senior citizen.

An office on the plaza in Santa Fe was a discreet welcome center for the professors who stepped off the Super Chief streamliner, blinking in the bright sunshine at the foot of the Sangre de Cristo Mountains. Among them was Otto Frisch, the Bach-whistling physicist who had sat on a log in Sweden and drawn the first pencil sketches of uranium shattering when hit by a neutron. He had also been instrumental in helping the British untangle the problem of separating U-235. Now he was given the opportunity to see his vision through to its inevitable violent end. Frisch had taken the ocean liner *Andes* over from Liverpool. Before boarding a westbound train in Virginia, the sight of a pile of oranges in a farmers' market sent him into a fit of laughter. He had not seen such fresh fruit for several years. Once he arrived in New Mexico, he could not contain his awe at being "among steep-walled canyons, accessible only by one rutted road; as isolated a place as one could wish for the most secret military establishment in the U.S.A." He received the customary greeting from J. Robert Oppenheimer: "Welcome to Los Alamos, who the devil are you?"

Most of those who arrived before Frisch had been welcomed with a series of blackboard lectures from one of Oppenheimer's favorite colleagues from Berkeley, Robert Serber, who laid out the task everyone faced in a series of chalked equations. The lectures were classified for years afterward, though they were legendary among those who heard them. Serber's opening words were the thesis of the entire Manhattan Project, and the culmination of an era in physics that had begun when a chemist accidentally left his uranium-coated photographic plates inside a drawer on a cloudy day in Paris.

"The object of the project," Serber began, "is to produce a potential military weapon in the form of a bomb in which energy is released by a fast neutron chain reaction in one or more of the materials known to show nuclear fission."

In other words, the point* was to vaporize Japanese lives and cities with uranium. In the midst of the lecture, Serber was passed a note asking him to please refrain from using the word *bomb*. He instead should use the more innocuous *gadget*. This would become the accepted Los Alamos euphemism.

The simplest way to build this gadget, Serber went on, was to machine-craft two different assemblies of pure uranium metal and then slam them together with great force. Within a fraction of a second, the scattering neutrons would trigger a chain reaction, ripping apart all at once. Adding a reflective tamper shield around the uranium would bounce the escaping neutrons back into the swarming beehive of the ultrahot core and boost the explosive power of the device. It would also decrease the likelihood of a fizzle and allow the team to get away with using less uranium, which was in preciously short supply. Serber included a crude sketch of a uranium metal slug being fired with a mini cannon into the curved receptacle of a receiving piece of uranium, as a penis enters a vagina. In the usual mechanic's vernacular, the convexity was termed the "male" part of the device, and the concavity the "female." (This is the most basic architecture of an atomic bomb, and the type of design experts say would most likely be used in a terrorist attack today.)

Otto Frisch started experimenting with assemblies of uranium that came within a hairsbreadth of becoming critical. He rigged up a small

---

* A postwar U.S. government assessment noted: "The expected military advantages of uranium bombs were far more spectacular than those of a uranium power plant."

tower that looked like an oil derrick and dropped a precisely measured plug of uranium down a central shaft so that it would slide briefly through a block of uranium with a corresponding hole cut in the middle, much as a firefighter passes through a hole sawed in the floor as he slides down a brass pole. The idea was to measure the neutrons that poured outward in the split second before gravity pulled the slug through the gap. The temperature rose several degrees in the room with each drop. The sardonic twenty-six-year-old physicist Richard Feynman likened this exercise to "tickling the tail of a sleeping dragon," and it became known on the mesa as the dragon experiment, or the guillotine. A vital exercise, but one that could have turned Los Alamos into a smoking ruin.

"Of course," said Frisch, "they quizzed me what would happen if the plug got stuck in the hole, but I managed to convince everybody that the elaborate precautions, including smooth guides and careful checks on the speed of each drop, would ensure complete safety. It was as near as we could possibly go towards starting an atomic explosion without actually being blown up, and the results were most satisfactory." As it turned out, the basic design of his dragon was a mimic of the gadget eventually dropped on Hiroshima.

But another of his experiments almost went horribly wrong. Frisch had been stacking blocks of enriched uranium without a reflective assembly—they were "naked," and so he called this the Lady Godiva experiment—when he leaned over the bricks to holler an order to a nearby graduate student. The red light on the neutron counters started glowing continuously, and Frisch realized what was happening: the white cloth of his lab coat and the water inside his chest were reflecting neutrons back into the uranium blocks. Frisch immediately knocked several of the blocks onto the floor with his forearm. After checking the meters, he saw he had already given himself a full day's allotment of radiation. Two more seconds and he would have been dead.

More than almost anybody else at Los Alamos, Otto Frisch had an intimacy with uranium. He had spent the last six years pondering its interior and envisioning its shape. The evocative term *fission*, suggesting a living being, had been his coinage. His pencil diagrams on the log in Sweden were the first crude etchings of the awesome powers coiled within uranium. There was a fearsome animal caged in this exotic metal, hot as the sun, but one whose instabilities could be accurately charted and precisely aimed. When the initial shipments of U-235 arrived at Los Alamos from the enrichment plant in Tennessee, Frisch had been among the first to hold the samples in his bare hands.

There had been a hundred or so blocks of the uranium, and they had gleamed like bright silver jewels in the box as an army sergeant looked on. Frisch had watched as the skin of the metal began to react with the oxygen in the dry Southwestern air, turning the blocks a sky blue color, and then purple. In their own way, Frisch had thought, they were beautiful.

"I had the urge to take one," he remembered. "As a paperweight, I told myself. A piece of the first uranium-235 metal ever made. It would have been a wonderful memento, a talking point in times to come."

But every speck of it had to be counted and hoarded; it was, at the time, the most valuable matter on the planet. Most of it was crafted into the protean sexualized parts that would constitute the heart of the first uranium bomb, code-named "Little Boy." The design was simple and elegant: When the bomb fell to a certain altitude (at about nineteen hundred feet for maximum destructive impact), a radar unit would close a relay and trigger a cordite explosion near the bomb's tail that sent a plug "bullet" of pure uranium down a steel barrel at a rate of 684 miles per hour into a series of uranium rings where the neutrons would shower uncontrollably. Frisch's "tickling the dragon" experiments had been invaluable in working out the calculations of speed and size. There would be no test of the U.S. uranium bomb before it was dropped; the science was reliable beyond a shadow of a doubt. There was also no more uranium metal to spare; every last ounce of it had gone into Little Boy, in which a total of 141 pounds was packed.

Its cousin, the plutonium bomb, was regarded as less of a sure bet. Its heart was a sphere of plutonium about the size of a softball encased in a cradle of high explosives designed to ignite at the same instant, their force amplified with lenses, pushing the plutonium into a small ball to cause a chain reaction. But if the explosives failed to fire at exactly the same instant—a moment measured in nanoseconds—the plutonium would be ejected and the bomb would fizzle. Its key similarity with the uranium bomb was that it also hinged on a simple act: The element had to be rammed into itself to achieve critical mass.

A test weapon code-named "Trinity" was taken out to a remote spot in the New Mexico desert to the west of the Oscura Mountains, a plain of dun-colored malpais called Jornada de Muerto, or "Journey of Death." At 5:29 in the morning on July 16, 1945, the plutonium device was set off, and it unleashed the equivalent of five thousand truckloads of dynamite within the space of a quarter second.

Otto Frisch was watching from twenty-five miles away.

By that time the first trace of dawn was in the sky. I got out of the car and listened to the countdown and when the last minute arrived I looked for my dark goggles but couldn't find them. So I sat on the ground in case the explosion blew me over, plugged my ears with my fingers, and looked the direction away from the explosion as I listened to the end of the count . . . five, four, three, two, one. . . . And then without a sound, the sun was shining; or so it looked. The sand hills at the edge of the desert were shimmering in a very bright light, almost colorless and shapeless. The light did not seem to change for a couple of seconds and then began to dim. I turned around but that object on the horizon that looked like a small sun was still too bright to look at. I kept blinking and trying to take looks, and after another ten seconds or so it had grown and dimmed into something more like a huge oil fire, with a structure that made it look a bit like a strawberry. . . . The bang came minutes later, quite loud though I had plugged my ears, and followed by a long rumble like heavy traffic very far away. I can still hear it.

The flash disintegrated the tower on which Trinity had been perched and turned the surrounding desert caliche into a lake of greenish glass nearly five thousand feet across. The physicist I. I. Rabi said, "There was an enormous ball of fire which grew and grew and it rolled as it grew; it went up in the air in yellow flashes and into scarlet and green. . . . A new thing had been born; a new control; a new understanding of man, which man had acquired

over nature." Rabi added, "It blasted; it pounced; it bored its way into you."
A representative of the Monsanto chemical company thought it looked like
"a giant brain, the convolutions of which were constantly changing."

Hardly a man given to poetry, Leslie Groves nevertheless reported to
the secretary of war a feeling of "profound awe" among nearly all those
who saw it. "I no longer consider the Pentagon a safe shelter from such a
blast," he noted in his July 18, 1945, memo.

This cold-blooded structural assessment, with a note of prolicide,
may have been the most eloquent thing the general was capable of
summoning: the knowledge that the building whose construction he supervised for sixteen months could be leveled in one second by this Thing
he had played a leading role in unleashing. It was, perhaps, his own way of
echoing the fragment that famously occurred to J. Robert Oppenheimer,
a line from the Hindu sacred text Bhagavad Gita: "I am become death,
the destroyer of worlds." At his side, the Harvard physicist Kenneth Bain-
bridge put it another way: "Now we are all sons of bitches." The man who
had discovered the neutron, James Chadwick, was silent as he watched the
mushroom cloud rise. Somebody clapped him on the back and he flinched,
making a choked sound of surprise, and was silent again.

The uranium bomb was shipped to an American air base on the island
of Tinian and loaded into the bay of the B-29 *Enola Gay** on the evening
of August 5, 1945. An airman thought it looked like "an elongated trash
can with wings." Only three members of the nine-man crew had been
told exactly what it was that they were scheduled to drop over Hiroshima
at 8:15 the following morning.

In New York City that evening, the phone rang in Edgar Sengier's
room at the Ambassador Hotel. He heard a male voice instruct him to
turn on the radio and keep listening. The caller then hung up without
identifying himself.

"I daresay they thought I had a right to know what was announced,"
Sengier said later.

Hiroshima had been founded on the delta of the Ota River in 1589

---

* Named for the pilot's mother.

during the samurai era and had grown to be an industrial city of about four hundred thousand people. It housed the headquarters of the Japanese Second Army. Until August 6, however, it had been spared the regular bombing suffered by other Japanese cities, so that the atomic bomb's full destructive potential would be on display. The epicenter was the Aioi Bridge in the middle of downtown, a place the pilot of the *Enola Gay* called the most perfect target he had ever seen.

The white flash ripped through the heart of town, spreading a nimbus of heat that reached five thousand degrees Fahrenheit. Birds exploded in flight. A squad of soldiers gazing up at the airplane felt their eyeballs melt and roll down their cheeks. People closest to the bridge were instantly reduced to lumps of ash. Farther out, people felt their skin burning and tearing and buildings disintegrated and streets boiled in their own tar. Children watched helplessly as their parents died underneath rubble.

"A woman with her jaw missing and her tongue hanging out of her mouth was wandering around the area of Shinsho-machi in the heavy black rain," reported one man. A junior college student said, "At the base of a bridge, inside a big cistern that had been dug out there, was a mother weeping and holding above her head a naked baby that was burned bright red all over its body, and another mother was crying and sobbing as she gave her burned breast to her baby."

A school for girls near the blast zone was vaporized, along with more than six hundred young students. Years later a memorial was erected on the site: a concrete female angel crowned with a wreath and holding a box with the legend $E = MC^2$.

In Washington, President Harry S. Truman released a prepared statement.

"The force from which the sun draws its power has been loosed against those who brought war to the Far East," he said. "Few know what they have been producing. They see great quantities of material going in and they see nothing coming out of these plants, for the physical size of the explosive charge is exceedingly small. We have spent two billion dollars on the greatest scientific gamble in history. We won."

He promised "a rain of ruin from the air, the like of which has never been seen on this earth" if the Japanese did not surrender immediately.

Less than three days later, the plutonium bomb was dropped on the city of Nagasaki, known as the San Francisco of Japan because of the beauty of its architecture and hilly seaside charm. It also had the largest population of Christian converts anywhere in the nation. Clouds had obscured the view of the city, and the plane had flown above it in circles, waiting for a gap to open. The bomb had to be dropped at the last minute, above a suburban Roman Catholic cathedral several miles from the original downtown target, a decision that accidentally saved the lives of thousands. The explosion was partly smothered by the hills, but it was still powerful enough to burn more than forty thousand people to death in the space of a few seconds. The amount of material inside the bomb that actually flashed into energy was but one gram—about one-third of the weight of a Lincoln penny.

An observer in the plane gazed at the mushroom cloud as the crew turned toward home.

"The boiling pillar of many colors could also be seen at that distance, a great mountain of jumbled rainbows in travail," he said. "Much living substance had gone into those rainbows."

4

APOCALYPSE

The world will end in fire. This belief can be found inside the official doctrine of a remarkable number of the world's religions.

In the twelfth century B.C., the prophet Zoroaster taught that a great confrontation between armies of good and evil would cause the mountains to collapse and turn into rivers of burning lava. The Aboriginals of Australia believe that mountains and plains are the frozen incarnations of the Creators, who will, if provoked, rise up and destroy all of existence with fire or flood. The Hopi of northern Arizona await the coming of a Fourth Age, and various prophecies describe a "gourd of ashes" thrown from heaven that will burn the land and boil the seas after young people reject the wisdom of their ancestors. Even the Hindu Bhagavad Gita, with its cyclical view of history, describes a "night of Brahman" when the universe will burst into fire before it is completely remade.

Norse mythology is unusually detailed: Three winters come with no summer in between them. The stars will vanish, the land will shake, men will kill their brothers, and the gods will make war with one another, culminating in a final orgy of violence. The Norse divinities have advance knowledge of their fate, called in German the Götterdämmerung, or "twilight of the gods," right down to who will kill whom with the slash of a sword, but they can do nothing to prevent it from happening.

A body of ancient Jewish writings makes reference to the *acharit hayamim*, or "the end of days." The most famous document was written shortly after the Jews were taken captive by the Babylonian king Nebuchadrezzar II in 606 B.C. A court adviser named Daniel dreamed of four beasts arising from the sea, the last with ten horns and iron teeth that "crushed and devoured its victims" before God appeared over a river of fire to initiate a kingdom with no end.

Some of Daniel's readers took this to mean that a massive battle would soon take place, perhaps near a hilltop outside the northern crossroads town of Megiddo. This was the spot where the beloved King Josiah was killed in battle just before the Babylonians drove the Israelites from their homes. The Har Megiddo (also called "Mount Megiddo" or "Armageddon") was the traditional Levantine geography of good's final triumph over evil and, with it, the purifying fire and end of history.

The Christian mystic John of Patmos had a dream about the end of the world in the first century A.D. He wrote an unforgettable account of a pale horse with Death as a rider; hail, fire, and a third of the earth burned up. "The sun turned black like sackcloth made of goat hair, the whole moon turned blood red, and the stars in the sky fell to earth, as late figs drop from a fig tree when shaken by a strong wind," he wrote. "The sky receded like a scroll, rolling up, and every mountain and island was removed from its place." The rain of fire and ruin must happen before the battle at Megiddo and the return of Jesus Christ to rule over all eternity. John's vision became known as the book of Revelation, the last book of the Bible. It is the most detailed account of the Christian apocalypse—in Greek, *apokalypsis,* which means to "uncover" or "disclose" the ultimate reality lurking behind the curtains of everyday color and sound.

Anticipation of the end did not dull with time. Saint Augustine found himself cajoling people in the third century to "relax your fingers and give them a rest" because too many were trying to calculate the last day. In the sixth century, Pope Gregory observed, "The world grows old and hoary and hastens to its approaching end." Viking attacks and comet appearances in the British Isles led many there to expect Christ's return in the year 1000; England's first land registry was called the *Domesday Boke—* that is, *Doomsday Book*—in anticipation of the final hour. The bubonic plagues of later centuries were seen as bowls of heavenly wrath poured out onto a wicked race, to be followed, mercifully, by the Last Judgment. The European exploration of the New World in the fifteenth century was accomplished in the light of a global sunset: Christopher Columbus noted that he sought to convert the natives of the West Indies to Christianity in a particular hurry because the world would soon be burning.

The end of time is not just a concept for the religious. The philosopher Georg Hegel taught that a dialectical progression of political forces would result in a new order; many of his admirers assumed that a final plateau would arrive within their lifetimes. Karl Marx foresaw an "end to history" when capital and labor joined in an egalitarian superstate; his followers tried to quicken that day through violent revolution. In the social upheaval of 1960s California, Joan Didion could write, "The city burning is Los Angeles's deepest image of itself. Nathanael West perceived that in *The Day of the Locust,** and at the time of the 1965 Watts riots, what struck the imagination most indelibly were the fires. For days one could drive the Harbor Freeway and see the city on fire, just as we had always known it would be in the end."

Is this a belief native to all of humanity? Is there a corner of the mind that is biologically predisposed to believe in an imminent end to the earth? The staggering number of religions and philosophies that forecast a burning before a final age of light makes it a question worth examining.

Carl Jung thought there was something archetypical about a global burning. "The four sinister horsemen, the threatening tumult of trumpets and the burning vials of wrath are still waiting," he wrote. Before him, Sigmund Freud viewed belief in Armageddon to be a common sign of schizophrenia. He also identified a dark urge he called destrudo, the opposite of libido. This is the destructive ecstasy in man, the flip side of his life-giving nature, the motor of warfare and slaughter. The images of the world going up in flames may be rooted in an individual's subconscious desire to be the one holding the torch. Or perhaps a belief in apocalypse is one of the mind's ways of subsuming the terrible foreknowledge of one's own death—apocalypse being a drama played out on a world stage that reflects the more mundane trauma of the individual self passing away. In *The Varieties of Religious Experience*, the philosopher William James

---

* One of the earliest and best novels about Hollywood. Its protagonist is the author of a screenplay called *The Burning of Los Angeles*.

wrote of faith as an essentially organic function; if that is true, then there may be no more profound organic stimulus to contemplate than death, which represents the end of existence as we understand it: the "burning of the earth" as a hazy and displaced amplification of the death pangs of the body before the final tide of white. Armageddon could turn out to be the most intimate of events—a Megiddo of the cortex.

But whether it stems from a genuine divine source or a neurological twinge (or both), the suspicion that the earth is ticking away its final hours has been salted throughout mythologies, religions, and cultures for many thousands of years. The suspicion of it exists on a grand collective scale in the same way that the narrator of Albert Camus's existential novel *The Stranger* perceives his own doom before his execution. "Throughout the whole of the absurd life I've lived, a dark wind had been rising toward me from somewhere deep in my future. . . ."

Historians call this idea endism and note that it tends to show up in fullest strength during times of crisis. And for those who witnessed the emergence of uranium bombs in 1945, the vocabulary of apocalypse came quite naturally.

The Harvard physicist George B. Kistiakowsky, invited to watch the Trinity explosion, called it "the nearest thing to Doomsday that one could possibly imagine. I am sure that at the end of the world—in the last millisecond—the last man will see what we have just seen." Normally sober newspapers reached for similar language. In the days immediately following the destruction of Hiroshima, the *Washington Post* called the uranium bomb a "contract with the devil" and concluded, "It will be seen that the life expectancy of our strange and perverse human race has dwindled immeasurably in the course of two brief weeks." The *Philadelphia Inquirer* termed it a "new beast of the apocalypse."

Even President Truman, who was famously coolheaded about the decision to use the weapon on Japan, wondered in his diary if the act he would soon authorize was "the fire destruction prophesied in the Euphrates Valley Era, after Noah and his fabulous ark."

Hiroshima has been called the exclamation point of the twentieth century, but it went much deeper than that. It threatened something embedded in the consciousness of the species: the imperative of collective survival. In the first week of August 1945, for the very first time in history, that sliver of the mind that watches for the end of the world was handed a scientific reason to believe the present generation may actually have been the one fated to experience the last burning, and that all the cities and monuments and music and literature and progeny and every eon-surviving achievement and legacy of man could be wiped away from the surface of the planet as a breeze wipes away pollution, leaving behind only a dead cinder, a tombstone to turn mindlessly around the sun. Forty thousand years of civilization destroyed in twelve hours.

For the religious and secular alike, uranium had become the mineral of apocalypse.

America went through a collective pause in the first days after Hiroshima, a period of quietude not unlike that after September 11, 2001, when the nation stopped cold for a week. Apprehension and confusion were widespread—a remarkable mood for a nation on the verge of winning a major war.

A correspondent for the *New York Sun* reported a "sense of oppression" in Washington as people talked about the new weapon. "For two days it has been an unusual thing to see a smile among the throngs that crowd the street." The sepulchral mood was noticed, too, by Edward R. Murrow of CBS News. "Seldom if ever has a war ended with such a sense of uncertainty and fear," he said, "with such a realization that the future is obscure and that survival is not assured." The president of Haverford College, Felix Morley, wrote, "Instead of the anticipated wave of nationalistic enthusiasm, the general reaction was one of unconcealed horror."

The *St. Louis Post-Dispatch* editorialized that the bomb may have "signed the mammalian world's death warrant and deeded an earth in ruins to the ants." The usually stolid *Corpus Christi Caller-Times* in south Texas concluded that "man's mechanical progress has outstripped his moral and cultural development," and addressed the new leveling force in the upper case. "Perfection of the Atomic bomb should make us readjust our values. . . . It should serve to give all of us a feeling of humanity."

When the news reached Los Alamos, there was a general excitement, and scientists rushed to book tables at Santa Fe's best restaurant to celebrate the achievement. But that night's party on the mesa was a grim affair. Almost nobody danced, and people sat in quiet conversation, discussing the damage reports on the other side of the world. When J. Robert Oppenheimer left the party, he saw one of his colleagues—cold sober—vomiting in the bushes.

"Certainly with such godlike power under man's control we face a frightening responsibility," wrote the military affairs reporter Hanson W. Baldwin in the *New York Times*. "Atomic energy may well lead to a bright new world in which man shares a common brotherhood, or we shall become—beneath the bombs and rockets—a world of troglodytes."

In an influential piece entitled "Modern Man Is Obsolete," written just four days after Hiroshima, the editor of the influential *Saturday Review*, Norman Cousins, tried to put a name to the dread:

Whatever elation there is in the world today because of final victory in the war is severely tempered by fear. It is a primitive fear; the fear of the unknown, the fear of forces man can neither channel nor comprehend. The fear is not new; in its classical form it is the fear of irrational death. But overnight it has become intensified, magnified. It has burst out of the subconscious and into the conscious, filling the mind with primordial apprehensions. . . . And now that the science of warfare has reached the point where it threatens the planet itself, is it possible that man is destined to return the earth to its aboriginal incandescent mass blazing at fifty million degrees?

Lurking behind much of the anxiety was an instinct of self-preservation. It was obvious that such a weapon could not remain exclusive to the United States forever and that what had been meted out to Hiroshima and Nagasaki could easily be returned to Chicago and Dallas, and in much deadlier portions. In an essay for the first postwar issue of *Time*, James Agee called Hiroshima "an event so much more enormous that, relative to it, the war itself shrank to minor significance. The knowledge of victory was as charged with sorrow and doubts, as with joy and gratitude."

Agee was even more pessimistic in a conversation with a friend: He called the bomb "the worst thing that ever happened" and predicted that it "pretty much guarantees universal annihilation." A poll taken one year later revealed that nearly two-thirds of Americans believed that atomic bombs would one day be used against the United States; an even higher percentage in another poll believed that most city dwellers would perish in such a war.

The fear of species extinction was not confined to America. In Rome, the *Vatican Press Bulletin* said the atomic bomb "made a deep impression in the Vatican, not so much for the use already made of the new death instrument as for the sinister shadow that the device casts on the future of humanity." In Britain, the *Guardian* observed caustically that "man is at last well on the way to the mastery of the means of destroying himself utterly." In France, the underground journalist Albert Camus, no stranger to combat, said, "Technological civilization has just reached its final degree of savagery. . . . Humanity is probably being given its last chance."

In Japan, the writer Yoko Ota, who had survived Hiroshima, remembered thinking the white flash was "the collapse of the earth which it was said would take place at the end of the world." Emperor Hirohito told his people he surrendered to prevent "the total extinction of human existence." In India the following year, Mahatma Gandhi said, "As far as I can see, the atomic bomb has deadened the finest feelings which have sustained mankind for ages." Also in India that year, the philosopher Yogananda reflected on the discovery of uranium. "The human mind can and must liberate within itself energies greater than those within stones and metals, lest the material atomic giant, newly unleashed, turn on the world in mindless destruction." In Russia, a biologist told *Pravda* that the Americans had plans to "wipe from the face of the earth . . . all that has been created through the centuries by the genius of mankind."

Eschatological thoughts had already occurred to some of the Manhattan Project scientists. Enrico Fermi took ironic wagers during the Trinity countdown that the resulting generation of heat would set the earth's atmosphere on fire and kill all life on the planet (this possibility had been raised and then quickly dismissed at the outset of the project). The day after Trinity, Leo Szilard persuaded sixty-seven fellow scientists at the University of Chicago to sign a confidential letter to Harry Truman urging him not to use the bomb on Japan. "If after the war a situation is allowed to develop in the world which permits rival powers to be in uncontrolled possession of this new means of destruction, the cities of the United States as well as the cities of other nations will be in continuous danger of sudden annihilation," he wrote. He would later refer to himself, and other atomic scientists, as "mass murderers."

The bomb's first inspirer, H. G. Wells, was still alive when the *Enola Gay* dropped its payload on Hiroshima. In the grip of liver cancer and too weak to write with much energy, he nevertheless managed a desultory essay for the *Sunday Express*, one of his last, in which he said that "there is no way out, around, or through the impasse" and "even unobservant people are betraying by fits and starts a certain wonder, a shrinking fugitive sense that something is happening so that life will never be the same."

Wells died at home in Regent's Park the following year; his ashes were scattered off the Isle of Wight.

There was a great deal of curiosity about uranium, a thing many Americans had never heard of before. Newspapers published diagrams of its enrichment cycle and maps pinpointing the now-unveiled complexes at Oak Ridge, Hanford, and Los Alamos. Graphics illustrating a chain reaction inside a cluster of U-235 were also displayed, albeit in generalized form. "This diagram merely illustrates the principle on which the atomic bomb works, not the specific processes occurring in the bombs dropped on Japan," noted *Time*, cautiously.

Worries arose over the security of uranium. Who owned it? Who would try to get it? Could it be stolen?

"Not only must this uranium be controlled, but, just in case a substitute is found, any suspect nations will have to be kept under surveillance to prevent the building of atom bomb plants," said the *Los Angeles Times* (an analysis that remains true today). At Rice University in Houston, the physics professor H. A. Wilson amplified the call for a multinational body to control the world's uranium supply "to see that the mastery of the destructive principle of atomic disintegration does not fall into the wrong hands." The comedian Bob Hope made it the punch line of a gloomy Valentine's Day joke: "Will you be my little geranium until we are both blown up by uranium?" A sketch artist on the boardwalk in Ocean City, New Jersey, said federal agents had interrogated him for several hours in 1943 because he drew an explosion and labeled it the work of a "ten-pound uranium bomb." The state geologist of Pennsylvania was moved to reassure local coal companies that they were still relevant—for the time being.

Such was the mystique accorded uranium in those days that *Scientific American* (apparently in all seriousness) proposed it be used as the world's monetary standard—not to be minted into coins, but to be used as a substance to guarantee the value of paper currency. Bars of uranium would play a role like that of bars of silver and gold in the nineteenth century. "Under such a scheme, atomic energy would be the basis of a reasonable currency whose value would be keyed to available energy, upon which depends production, the true measure of wealth," reasoned the magazine. The Federal Reserve Bank was not responsive to this idea.

There was curiosity about Shinkolobwe, the fabulous mine in Africa that had made all the difference. Edgar Sengier usually hated publicity, but did accept a congressional Medal of Merit from his friend Leslie Groves in a private ceremony. (When a new oxide of uranium was discovered in the Congo, geologists named it sengierite in his honor.) Sengier also granted an on-the-record interview in Paris to the newsman John Gunther in which he retold the story of slipping the uranium barrels out of Africa when nobody was watching. "I did this without telling *anything* to *anybody*!" he said.

Gunther later visited a town near Shinkolobwe and was allowed to see a piece of what he called the "brilliant, hideous ore."

"The chunk looked like a metal watermelon, pink and green, but it also had flaming veins of gamboge, lemon, and orange," he said. "The reflection was trite, but not difficult to summon—rocks like these have fire in them, not only figuratively but literally. The fate of civilization rests on a more slender thread than at anytime in history because of energies imprisoned in these flamboyant stones."

The mine itself was strictly off limits. Only one road led in or out. Arthur Gavshon of the Associated Press was turned away by armed guards at the gate. He later "met a blank wall of refusal" when he tried to talk to Union Minière officials. "We do not discuss uranium," one told him. The one thousand black workers who continued to labor in the pit for 20 cents a day had been instructed to keep silent, even though their work was no longer a wartime secret. Security had grown tight after reports that Soviet agents had set up a radio antenna in a nearby village and were recruiting some of the villagers for nonspecific "jobs."

One of the only visitors ever permitted inside the gates was the elderly Robert Rich Sharp, who had found the hill of radium as a young man almost forty years prior. Sharp had long since retired to a farmer's life in nearby Rhodesia, now Zimbabwe. He took a nostalgic trip back to his old haunts in the Congo in 1949, and Union Minière officials allowed him a brief honorary tour of Shinkolobwe so he could see what his discovery had wrought. The mine, he reported, was "surrounded by impenetrable barbed wire entanglements, with armed guards at the gates." But his memoirs discreetly make no further mention of what he saw there.

The caution seemed only logical, as "no metal in the world's history will be so jealously guarded or sought after," in the judgment of William L. Laurence of the *New York Times*. It had become the "most highly prized of all the natural elements, more precious than gold or any precious stone, more valuable than platinum, or even radium."

A respected science journal held up uranium as the new tool of global hegemony, equating this inanimate stone with the might of nineteenth-century armies. "If cannons were the final argument of kings, atomic power is the last word of great powers," said *Science News Letter* in its first commentary after the bombing. "This has apparently already happened without our realizing it in the case of the United States and the British Commonwealth. Whether we fancy it or not, these two great composite powers are now welded by a ring not of gold, but of uranium."

The mineral received more scornful treatment in one of the first pieces of fiction to incorporate the Hiroshima bombing as a plot point. The *Time* essayist James Agee, who believed uranium's ascendance was a guarantee of universal death, wrote his story "Dedication Day" in a blaze of anger. The story tells of a giant commemorative arch made of pure uranium metal about to be dedicated on the Mall in Washington, D.C. The arch bears the cryptic inscription THIS IS IT. The crowd is excited. But an atomic scientist makes a spectacle of committing suicide to atone for bringing such a monstrous thing into the world. The uranium, meanwhile, is "glistering more subtly than most jewels" in the capital's sunlight.

The story is a bricolage of ideas and images—a bit like *The World Set Free*, minus the optimism—and Agee lamented he couldn't find a way to adapt it into a movie.

But not all the initial reactions to Hiroshima touched on ominous themes, or on man's venality. *BusinessWeek* called it "that amazing atomic bomb." The *Las Vegas Review-Journal* hailed it as "one of the most important scientific achievements of all time" and speculated that mankind might be "on the threshold of one of those new eras which was ushered in by the invention of the steam engine, the internal combustion engine, harnessing electricity, discovering of the principle of radio...."

In the White House, Truman was confident the United States would enjoy a long exclusive on the atom, thanks in part to the false uranium forecast he had received from Groves. The stuff was supposed to be rare in the earth, and America had done outstanding work in securing most of the supply for itself—particularly at Shinkolobwe. Recent assessments had indicated the world's supply would last only until the year 2000. Truman also believed the enrichment process was too complex for the Russians to master. He said as much to J. Robert Oppenheimer during a conversation in the Oval Office.

"When will the Russians be able to build the bomb?" Truman asked.

"I don't know," said Oppenheimer.

"I know."

"When?"

"Never."

Oppenheimer went on to tell the president that some scientists felt they had blood on their hands for what had been accomplished at Hiroshima and Nagasaki, and for what the world could expect in the future.

An infuriated Truman would say later that he pulled out his hand-kerchief and handed it to the father of the A-bomb.

"Here," he said. "Would you like to wipe the blood off your hands?"

After Oppenheimer left the Oval Office, Truman turned to an aide and said, "I don't want to see that son of a bitch in here ever again!"

There was guilt and fear in America in the late summer of 1945, but there were also two important counterforces. The first was genuine patriotic pride. The country had just emerged victorious from a punishing two-front war with the help of a magic solution that had emerged from its innards. American muscle and intellect had forged this world-beating gadget and, in a flash over Japan, it had vaulted the country to the top of the international order.

The second was a hope that this new destructive force of uranium—the "Frankenstein," as one radio commentator termed it—might be turned into a servant of mankind. The beginning of the atomic age presented hazards, but it could bring a future of increased comfort and luxury. Managed carefully, it might even be a utopia. There could be atomic-powered cars, airplanes that would run on a pellet of uranium, ships that could fly to Mars in a week. In the words of an early newspaper report: "Furnaces of vest-pocket size. Power for whole cities produced from a few handfuls of matter."

These forecasts ranged from the ludicrous to the merely overoptimistic, but they helped ease the country through a period of disruption. They also highlighted a central truth about uranium that had been on display ever since Soddy's report first landed on the desk of H. G. Wells. Uranium was a mansion of physical violence, but the greatest part of its powers had always been rooted in the role it played in the human imagination.

In those first hours after Hiroshima, most of what America understood about atomic fission was based on newspaper reports and official statements of the U.S. War Department. And all were the literary output of just one man.

The person who would do more than anyone in history to present uranium as a friend to mankind held down two jobs at once—he was a beat reporter at the *New York Times* and also a paid author of press releases for the U.S. government.

William L. Laurence had already written the *Times*'s first stories about the discovery of atomic fission. He took a leave of absence in the crucial summer of 1945 to work as the "official journalist" for the Manhattan Project. Laurence was the only reporter permitted within the gates of Los Alamos and was allowed to personally witness both the Trinity and Nagasaki detonations. He had the unique role, therefore, of acting as a stenographer for the War Department while holding a position as the top science reporter for the nation's most influential newspaper.

The atom could scarcely have found a better spokesman. For Laurence, the advent of the atomic era was an unalloyed miracle: "an Eighth Day wonder, a sort of Second Coming of Christ yarn," as he once put it in a note to his editor. Most of the predictions he made were later discredited, but he succeeded in countering some of the fear of apocalypse by creating a sunny and blameless image around uranium. He bore a resemblance to that other influential poet of radioactivity, H. G. Wells, in that the two shared an exuberant prose style and a near-mystical appreciation for the powers of atomic physics. But though Laurence was working within the forms of nonfiction, he lacked Wells's sense of morality and balance. For him, the news was only good.

He was born in a Lithuanian village in 1888, a place he later described as "out of space and time," with mud streets and no running water. One of his earliest memories was grieving the death of a sick kitten. He went for a walk in the grain fields, asking, as he walked, "God, why did you kill my little kitten?" Field led to field, and the place he tried to reach—the spot "where the earth met the sky"—kept retreating in front of him. His earliest memory was also his earliest spiritual shock: the horizons never seemed to end. By the age of eight, he had memorized large portions of the Old Testament in the original Hebrew. But Laurence would later conclude that his prayers were useless and religion was a fairy tale.

Biblical rhythm and gravity would nevertheless have an influence on his development as a writer. So, too, would the experience of growing up in a repressive political system. His nose was permanently squashed from having been hit by the butt of a soldier's rifle. When he was seventeen, he threw bricks at policemen during an uprising against the Russian czar and was forced to flee. His mother smuggled him to Berlin inside a pickle barrel, and he eventually booked passage to America, home of the airplanes and radio that he had read about. Laurence had also read about the planet Mars and harbored a secret ambition to build an airplane capable of flying to the red planet, where he could perhaps learn "the secret to life." He told almost nobody about this boyhood dream until near the end of his life.

He painstakingly started to learn English at night, by comparing two translated versions of *Hamlet*. Shortly thereafter, he changed his name from Leid Sieu to William Laurence. His first name, he said later, was in honor of Shakespeare, and the second was for the peaceful suburban street in Roxbury, Massachusetts, where he settled. After winning an academic scholarship to Harvard, he studied both chemistry and drama and managed to land a job as a science writer at the *New York Times* in the days following the stock market crash of 1929. He stood out in the newsroom not only for what an editor called "an unquenchable, boyish enthusiasm for his job," but also for his deferential approach to scientists and the overcooked language he sometimes used to describe them. He was a short man with tall hair (a fellow reporter described him as "gnome-like") and an earnest but amiable demeanor that served him well in the environments in which he thrived. Laurence learned the journalist's trick of putting scientists at ease by asking an erudite question up front, letting the scientist believe that he was in the presence of a serious inquirer and not a dolt. This paid dividends: He wrote the first front-page story in *Times* history about a mathematical proof. In 1937, he shared in a Pulitzer Prize.

The defining day of his life may have been February 24, 1939, when he went uptown to hear an informal talk at Columbia University. Enrico Fermi and Niels Bohr had just revealed the possibility—by then, it was the inevitability—that a mass of uranium-235 could undergo a chain reaction if a neutron struck it. Laurence had already written about the discovery in an unbylined story for the *Times* headlined VAST ENERGY FREED BY URANIUM ATOM; HAILED AS EPOCH MAKING. The story had created a sensation, and Laurence wanted to follow up. He met the two in Room

403 of Pupin Hall, and the scientists chalked some sketches for him on the blackboard. He had heard the phrase *chain reaction* several times before, but this time it triggered in him a particularly vivid image, an epiphany not unlike the one experienced by Leo Szilard on a London street corner: a trillion trillion neutrons set loose in a nanosecond.

Laurence left the meeting in a daze. That night, he and his wife took their pet dachshund, Einstein, for a walk along the East River, underneath the stone footings of the Queensboro Bridge. Laurence was in a spooky mood, more remote than usual. The strange feeling from seeing that sketch on the blackboard was still with him. His wife, Florence, remembered him saying, as if in a dream, "A single bomb could destroy the heart of any city in the world. And the nation that gets it first may dominate the world."

From that point forward, Laurence dedicated his career to this one overarching topic; as he later put it, he became a "journalistic Paul Revere" in the name of the potential energy source within uranium. His enthusiasm was mingled with a dread that the Nazis would find the secret first. He recalled, years later, that "the world soon became for me one vast Poe-esque pit over which a uranium pendulum was slowly swinging down, while the victim remained unaware of his danger."

The flattering tone he reserved for theoretical physicists took on even more priestly coloring. At Harvard, Laurence had once aspired to write plays, and he maintained a lifelong membership in the Dramatists Guild of America. His taste for the theatrical—as well as, perhaps, a long-repressed sense of religious awe—now found full voice in his descriptions of the new field of atomic power. In a freelanced story for the *Saturday Evening Post* titled "The Atom Gives Up," he called Lise Meitner's accidental splitting of uranium "a cosmic fire" and "one of the greatest discoveries of the age," which would lead to "the Promised Land of Atomic Energy."

The article was not all puff: It contained a cogent description of fission before most scientists fully understood what was happening inside uranium. Laurence had a genuine talent for conceptualizing the more recondite elements of physics and through metaphor making them seem easy for his lay readers (though this, too, could go astray: He once described uranium as "an atomic golf course" on which professors were shooting balls of neutrons). He told his colleagues that the job of a science writer was to "take fire from the scientific Olympus, the laboratories and universities, and bring it down to the people." In a less grandiose moment, he talked of himself as a bee, moving pollen from flower to flower in order to "fertilize ideas." By 1940, reporters at the *Times* had started calling him "Atomic Bill." In the parlance of newsrooms, Bill Laurence had become a home-teamer, or a "homer"—one who had started to ape, consciously or not, the same language, mannerisms, and values of the people he covered.

The marriage became formal in the spring of 1945, when he received a visit from General Leslie Groves, who came to the third floor of the *New York Times* to make him a surprise offer. Laurence would be made a "special consultant" on a secret project then under way to harness exactly the same powers that he had been writing about. His primary job would be preparing the first U.S. government statements after the bomb was detonated over an enemy city. In return, he could have nearly unlimited access to the project on behalf of the *Times*, under the condition that Groves be given censorship power over the articles. They would also be stored in a military safe and stamped TOP SECRET until the end of the war. Fearing leaks and worried about the initial reactions of, as he called them, "crackpots, columnists, commentators, political aspirants, would-be authors, and world-savers" to the slaughter that was to come, Groves wanted a journalist with the credibility of the *Times* to shape America's first learnings about the bomb. It doubtlessly helped Laurence's case that he had a vision of uranium that approached the biblical. If his managing editor, Edwin James, had any misgivings about his science correspondent being on retainer to the War Department, they were not recorded. Laurence's dual-job status remained a secret both at the *Times* and to its readers until August 7, the day after the Hiroshima bombing.

"You will, for all intents and purposes, disappear off the face of the earth," Groves told him. That was just fine with Laurence, who later bragged that he was the one man in the country who knew so much about uranium that Groves was left with the option of either shooting him or hiring him.

Having surrendered his independence for the story of a lifetime, Laurence departed that spring for a private office at Oak Ridge, where the contents of his wastebasket were burned every night for security reasons. The men who collected these papers were, as he called them, "Tennessee hillbillies" who had been selected for the task because they couldn't read or write. Laurence's photo ID card gave him the same privilege and access as a colonel. He was taken on tours of Hanford and Los Alamos and introduced to the scientific team, many of whom he already knew from his coverage at the *Times*. He flew thirty-five thousand miles and had "seen things no human eye had ever seen before—that no human mind before our time could have conceived possible. I had watched in constant fascination as men worked with heaps of uranium and plutonium great enough to blow major cities out of existence." One day at Los Alamos, he had a close-up experience with the object of his fascination, much like that of Otto Frisch, who had wanted to pocket the first samples of pure uranium he was shown. Laurence wandered into the lab of Robert R. Wilson and found a pile of metal cubes on a table. He thought they looked like zinc.

"What's this?" he asked, casually picking one up.

"U-235," said Wilson, equally as nonchalant.

"I looked at the pile," recalled Laurence. "There was enough there to wipe out a city, but the fact that it was cut up into little cubes, separated here and there by neutron absorbers, kept the mass from becoming critical. . . . I hadn't believed there could be that much U-235 in existence."

All he could think to say at that moment was a banality—"My heavens!"—and Wilson quickly led him into the next room for a cup of tea.

Laurence was later allowed to see and touch Little Boy, the warhead into which this uranium would be packed, and it inspired him even more. "Being close to it and watching as it was fashioned into a living thing so exquisitely shaped that any sculptor would be proud to have created it, one somehow crossed the borderline between reality and non-reality and felt oneself in the presence of the supranatural," he wrote.

He had a seat on the Jornada de Muerto (a name that pleased him) when the Trinity bomb was set off. Laurence grabbed a pencil and started writing about it in the only way he knew—as a highly stylized news story, albeit one that could not be immediately published. His account sounds like a hymn, and in a way, it was. For Laurence, it was the nearest thing he experienced to religious ecstasy.

The atomic age began at exactly 5:30 mountain war time on the morning of July 16, 1945, on a stretch of semidesert land about fifty airline miles from Alamogordo, New Mexico, just a few minutes before dawn of a new day on that part of the earth. . . . The atomic flash in New Mexico came as a great affirmation to the prodigious labors of scientists during the past four years. It came as the affirmative answer to the, until then, unanswered question "Will it work?" With the flash came a delayed roll of mighty thunder, heard, just as the flash was seen, for hundreds of miles. The roar echoed and reverberated from the distant hills and the Sierra Oscura range nearby, sounding as though it came from some supramundane source as well as from the bowels of the earth. The hills said yes and the mountains chimed in yes. It was as if the earth had spoken and the suddenly iridescent clouds and sky had joined in one affirmative answer. Atomic energy—yes. It was like the grand finale of a mighty symphony of the elements, fascinating and terrifying, uplifting and crushing, ominous, devastating, full of great promise and great forebodings.

The Trinity blast, to Laurence, was "the first cry of a new-born world," which inspired some of the Nobel Prize winners on hand to dance and shout like pagans at a fertility rite. He compared the mushroom cloud to "a gigantic Statue of Liberty, its arm raised to the sky, symbolizing the birth of new freedom for man." He also observed: "It was as though the earth had opened and the skies had split. One felt as though one were present at the moment of creation when God said, 'Let there be light.' "

This last biblical invocation was among his favorites. Laurence never met a classical allusion that he didn't like, or attempt to employ. He drew frequently from Greek mythology, particularly the story of Prometheus, the renegade god who brought fire down to earth. Another preferred trope was the scientist as Moses, leading all of civilization to a "land of milk and honey" where Arctic snow was melted before it could touch the ground and disease was a hobby of the past. Metaphors from the Pentateuch came naturally to him, but one that he conspicuously neglected would become a favorite of later critics of the atomic program: that of the tree of knowledge, where man ate and was sorry for it.

Among science writers, Laurence may well have been the worst prose stylist of his generation. His writing was arid and chilly, even though it was packed with more flourishes than a romance novel. Uranium was to Laurence, at various points, "a cosmic treasure house" and a "philosopher's stone" or a "Goose that laid Golden Eggs," which "brought a new kind of fire" that led to "the fabled seven golden cities of Cibola."

These messianic word-pictures of a life to come, though wildly overoptimistic, helped create in the American public a generally positive and hopeful feeling about the dawn of the new atomic age. They also helped to blunt the disquieting sense that the world was about to enter its last days. To the contrary, said Laurence; earth was about to become a paradise. "Man had found a way to create an atmosphere of neutrons, in which he could build an atomic fire more powerful than any fire he had built on earth. With it he could create a new civilization, transform the earth into a paradise of plenty, abolish poverty and disease and return to the Eden he had lost." Faith in God had begun to desert Laurence when his little white kitten died, but his faith in the rational powers of science was unshakable. H. G. Wells could not have created a character more thoroughly sold.

The major disappointment of his war experience was missing the Hiroshima explosion. His flight from San Francisco to the Pacific island of Tinian, the final staging area for the mission, was delayed because of weather, and he arrived in time for only the final briefing, too late to get a seat on the bomber. But he played a major role nonetheless. He wrote the U.S. government's announcement of the atomic bomb, delivered by President Harry S. Truman on August 6, 1945.

"Mine has been the honor, unique in the history of journalism, of preparing the War Department's official press release for worldwide distribution," Laurence said later. "No greater honor could have come to any newspaperman, or anyone else for that matter."

An early draft had been rejected. Laurence had been in full rhetorical flower, describing a "new promised land of wealth, health and happiness for all mankind," but an adviser to the president found it "highly exaggerated, even phony," and the White House opted for a more sober version in which Truman identified the bomb as merely harnessing "the basic power of the universe." Laurence's articles about the inner workings of the Manhattan Project were handed out that same day to newspapers both in America and abroad. In most cases, they were simply run verbatim. The public devoured the news about "Mankind's successful transition to a new age, an Atomic Age." Almost every fact and coloring the world absorbed in those first hours had come from the pen of Bill Laurence.

Near the end of his life, he recalled his flight toward Nagasaki as the sole journalistic observer of the second atomic bomb drop—an event he later described as the culmination of his career. He was sitting in his seat in the B-29, gazing out over a pure "endless stretch of white clouds" over the Japanese islands. At the time, the target city was Kokura, which was changed at the last minute to neighboring Nagasaki due to inclement weather. Kokura would be saved by a cloud pattern. Laurence nonetheless contemplated its destruction from an Olympian loftiness, with a certain amount of mad joy. He said:

> It was early morning; it was dark and I was thinking of the town of Kokura being asleep and all the inhabitants having gone to bed, men, women, and children . . . they were like a fatted calf, you know, saved for the slaughter. . . . And here I am. I am destiny. I know. They don't know. But I know this was their last night on earth. I felt that the likelihood was that Kokura would be completely wiped off the face of the earth. I was thinking: There's the feeling of a human being, a mere mortal, a newspaper man by profession, suddenly has the knowledge which has been given to him, a sense—you might say—of divinity.

Laurence had taken the final emotional step of merging his self-consciousness not just with the nuclear weapon ("I am destiny") riding underneath him but with the godlike powers that it held inside. He could say, without apparent embarrassment, that the sense of being dissolved into the unstable heart of uranium was a feeling akin to that of a rhapsodic worshipper becoming united with the divine through prayer or song. The psychologist Robert Jay Lifton, writing with Greg Mitchell, identified this as Laurence's moment of "immortality hunger" through his association with the bomb. Fission had become his religion.

The news story he wrote, scribbled on a notepad on his knee, was more restrained in its apocalyptic tone, though no less colorful. The cloud of fire he saw blossoming over the city "was a living thing, a new species of being, born right before our incredulous eyes . . . creamy white outside, rose-colored inside." Up to seventy thousand people were incinerated within. The eyewitness account did not run in the *Times* until nearly a month afterward because of wartime censorship. A reader wrote to say it was the finest descriptive passage outside of the writings of Edgar Allan Poe; Laurence was proud of the comparison and mentioned it in his later years.

He was not allowed to visit the wreckage of either Hiroshima or Nagasaki to see the effects of what he had touted, having been called home to New York to write a ten-part series for the *Times* on the creation of the atomic bomb. Laurence did find time, however, to write a lengthy story on September 12 debunking reports that the bomb blasts were associated with deadly levels of gamma rays—a cold scientific fact he almost certainly knew to be true, according to the journalists Amy Goodman and David Goodman.

This public relations crisis was sparked after an Australian reporter named Wilfred Burchett had already dodged a cordon and snuck into the ruins of Hiroshima, where he found thousands of people dying from an unknown wasting disease. His story in the London *Daily Express* began with these words: "In Hiroshima, thirty days after the first atomic bomb destroyed the city and shook the world, people are still dying, mysteriously and horribly—people who were uninjured in the cataclysm from an unknown something which I can only describe as an atomic plague." Burchett interviewed Japanese doctors who described what would soon be known as the classic symptoms of radiation sickness: teeth and hair falling out, appetite loss, bleeding from the nose and mouth, festering burns. American officers expelled Burchett from Japan and confiscated all his photographs, but they failed to stop the story from running in Britain.

The U.S. military moved quickly to squelch all news of radioactivity. There were worries in the Pentagon that the bomb would be compared to German mustard gas in World War I, or other types of wartime atrocities. Laurence was taken on a staged tour of the Trinity site, where he was shown Geiger counters "proving" that no appreciable levels of radiation remained in the area. This was a ridiculous exercise, as the devices would have said nothing about the deadly levels present during the first seconds of the blast. In the first paragraph of a front-page story headlined, sneeringly, U.S. ATOM BOMB SITE BELIES TOKYO TALES, Laurence praised this "most effective answer" to "Japanese propaganda." He quoted Groves as attributing almost every single Japanese death to fire and blast damage. If radioactivity was so dangerous, wondered Groves, then why was grass now growing on the Hiroshima parade grounds?

The premise of the story was, of course, grossly incorrect. An estimated seventy thousand people were killed in the initial attack on Hiroshima, but the final body count was closer to one hundred thousand as radiation sickness and, eventually, cancer took their toll. Laurence was in an excellent position to know about the radiation, thanks to his level of scientific access to Los Alamos and to the measurable levels of gamma rays in the air immediately after Trinity—enough to kill rabbits far away from the epicenter and to create, as he later admitted, gray ulcers on the hides of faraway cows.

The Associated Press was already running speculation that the "uncanny effects" on the Japanese were due to gamma rays—an entirely correct assessment. But this would not have been the first time that Laurence introduced deliberate falsehood into the record. Before the Trinity test, Groves had ordered him to prepare a statement blaming the flash in the desert on the accidental ignition of an ammo dump. The local New Mexico newspapers printed the bogus story without question. "The secret had to be kept at all costs and so a plausible tale had to be ready for immediate release," Laurence explained later, without a trace of regret.

Laurence's series on the making of the atomic bomb, including his eye-witness account of the Nagasaki explosion, won a Pulitzer Prize. A later memoir, titled *Dawn Over Zero* and written with the same breathless moxie of his news style, became an international bestseller. He was promoted to science editor and took an office on the *Times*'s coveted tenth floor. At one point after the war, he had the unique honor of learning that FBI agents had been sent out to public libraries in 1940 to remove all copies of the edition of the *Saturday Evening Post* that carried his story on uranium. A translated copy of the same story, wrapped in cellophane, was discovered after the war, inside the safe of a German physicist assigned to work on the hapless Nazi weapons program. Even the enemy had apparently considered his writings prophetic. Until he died of a blood clot while vacationing in Spain in 1977, Laurence was treated as an authority on nuclear energy and as a minor celebrity. His obituary in the *Times* noted: "He occupied honored places on daises at major affairs of scientific and other organizations, and his short, chunky frame frequently stood on the lecture platform."

Laurence was not completely blind to the moral ambiguities raised by the dawn of the age that he had trumpeted so loudly. His descriptions of Nagasaki made reference to the "living substance" boiling within the rainbow cloud. He acknowledged the "great forebodings" of the power now under America's control. It was he who had extracted and publicized that ominous reverie from J. Robert Oppenheimer: "I am become death, the destroyer of worlds." And in a 1948 article for the magazine *Woman's Home Companion*, he allowed that the future with uranium could be a hell instead of a heaven if it were mishandled. The image he used was by that time already shopworn—nearly as much as "the genie is out of the bottle"—but it was still an admission of an alternative to nuclear Eden. Laurence being Laurence, he made it stark.

"Today we are standing at a major crossroads," he wrote. "One fork of the road has a signpost inscribed with the word *Paradise*, the other fork has a signpost bearing the word *Doomsday*."

The country began to accept Armageddon as a possibility, but one rel-egated to the future. The United States, after all, was still the only nation to possess atomic weapons, and the newspapers were quick to note that most of the uranium inside the earth was now under direct American control. Experts also reassured the public that the technological abilities of other countries—especially the Russians—were hopelessly backward. The celebrations that followed Japan's surrender eight days after Hiroshima also helped lighten the mood. There was time to relax.

The word *atomic* soon became an adjective for all things mighty and exotic. Cleaning services and diners renamed themselves with the neologism. A set of salt-and-pepper shakers in the shapes of Fat Man and Little Boy—the bombs dropped on Nagasaki and Hiroshima—were popular sellers. A few enterprising bartenders concocted strange-tasting drinks and called them "atomic cocktails," invariably followed up with a call to the local newspaper in hopes of a write-up. The Hotel Last Frontier blended vodka and ginger beer; "a perfect toast to victory," enthused the *Las Vegas Review-Journal*. At the Washington Press Club, the recipe was Pernod and gin. Yet other versions mixed aquavit and beer; gin and grapefruit juice; vodka, brandy, and champagne. The proliferations inspired the jazz vocalist Slim Gaillard to write a new song.

> It's the drink you don't pour
> Now when you take one sip you won't need anymore
> You're as small as a beetle or big as a whale
> BOOM!
> Atomic Cocktail

From the corpse of the Manhattan Project, a new government agency was formed: the Atomic Energy Commission. Its director, David Lilienthal, flew around the country making speeches to civic groups and schools. At a dramatic point in his address, he would pull a lump of coal from his pocket. A piece of uranium this big, he would say, could keep Minneapolis warm for a whole winter. Utopian projections such as this led the *New Yorker* magazine to propose that the "pea" be adopted as the new standard unit for the measurement of energy. Lilienthal was privately skeptical about the utopian promises for atomic power, but in public he was as irrepressible as William L. Laurence. He helped stage a

Walt Disney–style exhibit in New York's Central Park called Man and the Atom, featuring displays sponsored by the government's biggest atomic contractors. One of the handouts from Westinghouse was a slick comic book called *Dagwood Splits the Atom*, in which the goofy husband from the strip *Blondie* blows a neutron through a straw and shatters a fat, puffy nucleus of U-235. "General Groves himself, in yet another of the high-level policy decisions of his career, chose Dagwood as the central character," noted the historian Paul Boyer.

Lilienthal's agency had the contradictory—some said impossible—mission of encouraging peaceful atomic energy while also gathering uranium for military use. One of its important early choices was picking a suitably useless piece of land as a spot to test new weapon prototypes. A stretch of basin and range one hour north of Las Vegas was sealed off and renamed the Nevada Proving Ground. This was the badlands described by the explorer John C. Frémont a hundred years before as "more Asiatic than American in its character"—a waterless sinkhole of brush and weeds hemmed in by peaks of basalt and granite. The Army Air Corps had first used it for target drills, and one official explained, "The land was cheap because it really wasn't much good for anything but gunnery practice—you could bomb it into oblivion and never notice the difference." More than one hundred atomic bombs would be detonated there between 1951 and 1963, and hundreds more would be set off inside tunnels below the desert floor.

The surface blasts were plainly visible from Las Vegas and made for spectacular viewing. AEC officials made repeated assurances that the tests were perfectly safe, and the city embraced its new role as "The A-Bomb Capital of the West" with characteristic brio. A stylist at the Flamingo hotel created a hairdo in the shape of a mushroom cloud; it involved wire mesh and silver sprinkles and cost $75. A motel renamed itself the Atomic View and told its guests they could watch the detonations from poolside chairs. Dio's Supermarket on 5th Street boasted of prices that were "Atomic—in the sense of being small." Roulette balls and craps dice were occasionally nudged by blasts; casinos posted signs that made such events subject to the ruling of the pit boss. The Las Vegas High School class of 1951 adopted the mushroom cloud as its official mascot and painted one outside the entrance to the school. Clark County redesigned its official seal to include the same logo. "Dazzled by atomic eye candy," wrote the Nevada historian Dina Titus, "citizens were virtually hypnotized into acceptance." A local housewife named Violet Keppers was taken out to the viewing stands to witness a blast, and she later wrote about it for *Parade* magazine. "All my life I'll remember the atomic cloud drifting on the wind after the blast," she wrote. "It looked like a stairway to hell."

The first indication that anything was wrong came after a May 19, 1953, explosion—later nicknamed "Dirty Harry"—in which the wind had shifted unexpectedly. Ranchers in neighboring Utah watched with puzzlement, and then anger, as herds of their lambs and ewes fell sick and died. Under their wool, many were showing ugly running sores similar to those seen by William L. Laurence on cattle far away from the Trinity blast. The AEC told them their livestock had died of malnutrition and cold weather. But on one hard-hit ranch, an official with a Geiger counter was overheard hollering to his companion, "This sheep is as hot as a two-dollar pistol!" The ranchers sued for damages and lost.

The radioactive clouds had scattered dust over the nearby town of St. George, Utah, where, the following year, John Wayne and Susan Hayward would spend three months in a canyon filming *The Conqueror*, an epic about Genghis Khan (today generally regarded as Wayne's worst movie). The canyon was breezy, and the cast and crew were constantly spitting dust from their mouths and wiping it from their eyes. Almost half of them, including Wayne and Hayward, would eventually die from assorted cancers, a rate three times above the norm. The downwind plume from the test site, which generally blew to the north and east, proved to be an accurate map for later elevated thyroid cancer occurrences. The number of people sent to early graves is still a matter of controversy; most estimates put the count at well more than ten thousand, spread across five mountain states.

But the testing was not about to stop. When a state senator called for a ban, he was vilified by the local newspapers. An AEC board member said, "People have got to live with the facts of life and part of the facts of life are [sic] fallout." There were repeated calls for Nevada to embrace the explosions as a symbol of the continuing American pioneer spirit. Most residents also welcomed the high-wage federal jobs and the prestige. "More power to the AEC and its atomic detonations," cheered the *Las Vegas Review-Journal.* "We in Clark County who are closest to the shots aren't even batting an eye." The new weapons were part of a grand strategy to grow the U.S. arsenal to deter a perceived threat from the USSR—a decision to swell, rather than shrink, the American dependence on uranium as a centerpiece of its defense posture.

But this decision came only after a serious discussion about creating an international body to safeguard the world's supply of uranium and nuclear weapons—a scenario straight from *The World Set Free.*

President Truman had asked both Lilienthal and Undersecretary of State Dean Acheson to write up a proposal for a powerful multinational regime to purchase St. Joachimsthal and Shinkolobwe and all new mines, destroy all the bombs the United States had already manufactured, and assume guardianship of all the atomic facilities in every nation. Atomic power would be encouraged for peaceful use, leading eventually to total disarmament and the obsolescence of war. That such a dovish proposal could have ever come from the White House of Harry Truman, even when he believed the Russians were incapable of building the bomb, is testament to the millennial panic and social readjustment that had swept through the country after Hiroshima—as well as to the influence of newspaper prophets such as "Atomic Bill" Laurence.

A revised outline was presented to the first meeting of the United Nations Atomic Energy Commission on June 14, 1946, in a speech by the silver-haired financier Bernard Baruch. He began in high Laurentine style:

> We are here to make a choice between the quick and the dead. That is our business. Behind the black portent of the new atomic age lies a hope which, seized upon with faith, can work our salvation. If we fail, then we have damned every man to be the slave of fear. Let us not deceive ourselves: we must elect world peace or world destruction.

The White House already knew that the plan would be dead on arrival. The USSR objected to the idea of giving up its uranium deposits and enrichment facilities, both of which it was frantically developing in secret. The United States also continued to manufacture weapons even while the plan was under debate, which gave the Soviet UN delegate Andrei Gromyko license to accuse his negotiating partners of hypocrisy. Baruch eventually resigned from the commission, and the proposal died quietly several months later.

This ploy came at a time when the geopolitical hierarchy was in the midst of a shift not seen since the defeat of Napoleon Bonaparte. Britain was in full retreat from its empire, granting independence to India, Pakistan, Israel, and dozens of other territories. The USSR was aggressively pushing its mandate in Eastern Europe and backing Communist rump governments and guerrilla fighters in places as disparate as Greece and North Korea. The American possession of atomic bombs, and the infrastructure it took to make them, were seen as a strategic asset in an unstable world—the ultimate trump card. Uranium did not march like an army: It was a silver scythe that could decapitate a nation in an hour. As Leslie Groves concluded in a memo, "If there are to be atomic weapons in the world, we must have the best, the biggest, and the most."

American nuclear thinking was thereby solidified for the next forty years, crystallized in an elaborate paradox that sounded like a geometric proof. There could be no defending against a nuclear strike. But such an attack would invite retaliation, eliminating both nations. No sane leader would trigger such a thing. Therefore the devices most able to vaporize the enemy were also useful for ensuring that such a thing would never occur.

The warheads might just as well have been made of cardboard—it was their abstract threat that counted. As it was in H. G. Wells's time, uranium's physical powers were far secondary to the power of the narrative that man could craft around them. The United States and the USSR would be locked together in this crude, but effective, story for almost forty years. J. Robert Oppenhemier would memorably call the standoff "two scorpions in a bottle, each capable of killing the other but only at the risk of his own life."

This psychological doctrine, first called massive retaliation and later known as mutually assured destruction, or MAD, was originally formulated by a young instructor at Yale named Bernard Brodie, who published the influential book *The Absolute Weapon: Atomic Power and World Order* in 1946. "Thus far the chief purpose of our military establishment has been to win wars. From now on its chief purpose must be to avert them," he wrote in the book's most quoted passage. Suspicion of one's rival was a healthy thing, argued Brodie. That was, in fact, the basis on which a future peace would be secured. Deterrence was best achieved through preparing for war.

This was the intellectual cornerstone of the arms race. If war did come, it would be, in the popular phrase of the day, a "push-button war." By the beginning of 1953, the United States had about a thousand nuclear weapons in its arsenal. Less than ten years later, there were twenty-one hundred, with a grandiose air, land, and submarine deployment pattern—the "triad"—and enough firepower to level Hiroshima again an estimated million and a half times. Pentagon officials struggled to find Russian targets to match the bombs and not the other way around. The buildup eventually cost taxpayers an average of $98 billion a year for weapons that were effectively useless except as public relations tools. "If you go on with this nuclear arms race," warned Winston Churchill in 1954, "all you are going to do is make the rubble bounce." The warning went unheeded. This ring of uranium would eventually cost the United States more than $10 trillion in armaments and support; the historian Richard Rhodes has pointed out that this figure is larger than the entire economic output of the United States in the nineteenth century.

MAD had other weaknesses. Atomic strength would prove useless in regional conflicts where the United States had an interest. In fact, it was worse than useless. The bombs sucked away money and attention from a conventional fighting apparatus, and tended to encourage magic-bullet thinking. President Eisenhower's National Security Council had pressed him in 1954 to use nuclear weapons on the Vietnamese insurgents who had a hapless French colonial garrison surrounded at Dien Bien Phu. "You boys must be crazy," he told them. "We can't use those awful things against the Asians for the second time in ten years. My God."

There may have been another, more distinctly human purpose for the United States to have sped up its pursuit of a uranium-based defense. Institutional momentum is a hard—if not impossible—thing to stop. Careers and budgets were now at stake. The American military had been slow to awaken to the possibilities of uranium during the war—there was that year-long wait after Einstein's letter—but now a $2 billion assembly line spread across three states and two continents was humming at warp speed. It was the showcase for the ingenuity and prestige of the nation, and the backbone of what William L. Laurence and others had hailed as the most important scientific advance of all time.

To have just locked the doors of the Manhattan Project and walked back toward the banality of nitrate bombs would have been a denial of one of man's basic urges, that of creation and discovery. "Nuclear explosions have a glitter more seductive than gold to those who play with them," said the physicist Freeman Dyson. "To command nature to release in a pint pot the energy that fuels the stars, to lift by pure thought a million tons of rock into the sky, these are exercises of the human will that produce an illusion of illimitable power." As the writer Rebecca Solnit has observed, a nuclear test gives man the power to make a star, if only for a moment. Now a huge federal apparatus was pushing this Dionysian urge, with a lavish budget and patriotic fervor behind it.

The apocalyptic fears of postwar America never completely disappeared. When he accepted the Nobel Prize in literature in 1950, the Mississippi novelist William Faulkner could declare with authority, "There are no longer problems of the spirit. There is only one question: When will I be blown up?" The specter of death had become just another part of urban living, said *Time* magazine, noting that the modern man on the sidewalk now kept watch for "the blinding flash of a terrible light, brighter than a hundred suns." The essayist E. B. White beheld a grim epiphany about his beloved New York City, and a modern reader might see shadows of September 11, 2001, in his thoughts: "A single flight of planes no bigger than a wedge of geese can quickly end this island fantasy, burn the towers, crumble the bridges, turn the underground passages into lethal chambers, cremate the millions. . . . In the mind of whatever perverted dreamer might loose the lightning, New York must hold a steady, irresistible charm."

The civil defense programs of the 1950s tried to soften this image with a comforting new story: An atomic war did not have to mean instant death. With the right precautions, said the experts, it could be survivable. A paperback guide called *How to Survive an Atomic Attack*, written by a Pentagon consultant, advised citizens to wear long pants, a hat, and—if possible—rubber boots to protect against fallout.

"Just keep the facts in mind and forget the fairy stories," it advised. "Follow the safety rules. Avoid panic. And you'll come through alright."

The cellars of thousands of schools and libraries and suburban malls were classified as fallout shelters. Metal signs colored yellow and black were bolted up on exterior walls to mark their presence. Stocks of food products, many of which grew moldy, were stored inside. President Eisenhower ensured that the interstate highway system received generous funding: It was envisioned as a national web of blacktop to evacuate cities in case of attack. The highways emanating from city cores quickly became the spines of new suburbs, where many said they felt safer from crime and possible nuclear attack. At the U.S. General Services Administration, Tracy Augur concluded that the traditional city was obsolete and "not only fails to offer any security against enemy attack; it actually invites it, and places the lives and property of the citizens in jeopardy." He also noted that some cities "seem to feel as if they don't rate unless they contain at least one good A-bomb target." A suburban advocate named Peter J. Cunningham wrote to the *Chicago Daily Tribune:* "Leading scientists declare that the secret of the atomic bomb cannot be controlled; therefore it behooves us to spread out over the countryside so that such bombs may fall without killing so many people at a time, or causing so much damage." The editors headlined his letter PREPARING FOR ARMAGEDDON.

This thinking was endorsed at top levels. President John F. Kennedy was an enthusiastic proponent of fallout shelters and had one built at the family compound in Palm Beach, Florida. The White House authorized the hollowing of a basalt mountain in the horse country near Mount Weather, Virginia, to house a replacement government in case Washington, D.C., should be lost. The command center is said to contain a cafeteria, a hospital, bunk rooms, a crematorium for the disposal of dead members of the executive branch, and a television studio for postbellum broadcasts. (The facility still gets used; by some accounts, this was the "undisclosed location" where Vice President Dick Cheney was relocated after the September 11 attacks.) Congress retaliated by building its own secret bunker underneath the Greenbrier Hotel in White Sulphur Springs, West Virginia, in the name of "continuity of government."

Those not lucky enough to hold elective office were encouraged by the Office of Civil Defense and Mobilization to build their own fallout shelters and stock them with food, blankets, flashlights, and Geiger counters. Approximately one million families heeded the call and constructed subterranean refuges that ranged from a shovel-dug hole in the backyard covered with cardboard to plush accommodations that cost as much as a new house and doubled as a rumpus room.

American children were the target audience for instructional books such as Walt Disney's *Our Friend the Atom*, which deployed the most fundamental of nuclear clichés: a genie released from his bottle who has the power to kill his master. "We have the scientific knowledge to turn the genie's might to peaceful and useful channels," concluded the introduction. At the same time, children were taught to watch for any sign of a white flash and were subject to classroom drills in which everybody from the principal on down was supposed to hide underneath his or her desk. A cartoon instructional film, produced by New York's Archer Films and sponsored by the U.S. government, featured a turtle whose shell protected him from radiation, just as a desk was supposed to shield a child. The signature jingle was widely ridiculed.

> There was a turtle by the name of Bert
> And Bert the Turtle was always alert
> When danger threatened him he never got hurt
> He knew just what to do
> Duck and Cover

The Atomic Energy Commission also tried its best to put a friendlier face on uranium by emphasizing its peaceful uses. "We were grimly determined to prove that this discovery was not just a weapon," said David Lilienthal. This was despite the growing consensus among scientists that the "miracles" of atomic energy were nothing more than outright fairy tales. The director of research for General Electric reported that "loud guffaws" could be heard in the laboratory whenever somebody repeated a Laurentine promise of a future without disease, toil, or famine. "The economics of atomic power are not attractive at present, nor are they likely to be for a long time in the future," he said. "This is expensive power, not cheap power, as the public has been led to believe."

But could the weapons themselves be used for something other than apocalypse? Testing at the Nevada Proving Ground had revealed that a nuclear bomb buried in a deep shaft underneath a mountain would vaporize the surrounding rock and make a huge cathedral-like space inside the earth, ablaze with radioactivity. But the only noticeable effect aboveground was that the mountain bucked approximately six inches at the moment of the blast. This led to speculation on the part of scientists at the Livermore Radiation Laboratory in California: Why think of the uranium chain reaction as just a tool for flattening enemy cities? Why not bury one inside a rocky shoreline and use it to instantly carve out a harbor? Or tear a canal across Israel that would rival the Suez? Or perhaps set one off underground in a region of oil sands to release trapped petroleum reserves?

This was the mentality behind the Plowshare program, an ill-fated AEC initiative that explored the use of nuclear bombs as construction tools. The Atchison, Topeka & Santa Fe railroad went so far as to consult with the state of California about blasting a two-mile channel through the Bristol Mountains for a speedier route from Flagstaff, Arizona, to Los Angeles. Plowshares did a feasibility study on setting off a string of hundreds of bombs simultaneously to widen the Panama Canal. Helpfully included in the study was a map of the likely patterns of radioactive fallout that would blanket large portions of Central America.

Commissioner Willard Libby insisted that it would be unnatural *not* to use nuclear weapons in such a fashion. "There is a natural law, I think, which requires us not to turn our backs on nature," he told a congressional committee. (By "nature," he meant nuclear weapons.)

The program was inaugurated with Project Gnome, a scheme to detonate a small device, with approximately one-seventh of the yield of the Hiroshima bomb, inside a siltstone aquifer under a piece of barren scrubland outside Carlsbad, New Mexico. The idea was to instantly boil the groundwater into steam to see if power generation might be feasible.

Reporters and scientists were invited to watch—actually, listen—from a reviewing stand on December 10, 1961. There was a muffled bang, and the ground heaved and punched upward, cracking a wellhead. Almost immediately, radioactive steam and black smoke began pouring out. "The blast burst through a cavern shaft, ignited a chemical charge prematurely, and jolted observers five miles away," noted one correspondent. It also kicked up a giant cloud of radioactive dust that drifted across a nearby highway. When workers tunneled in more than half a year later to inspect the damage, they found a hollow chamber about the size of the U.S. Capitol dome. The rock walls were colored brilliant shades of blue, green, and purple and bore an angry surface temperature of 140 degrees Fahrenheit. Drilling at the site is prohibited today; the radiation still poses a danger. The Plowshare program went on to detonate two dozen more warheads, mostly underneath the Nevada Test Site, before finally being discredited.

The Gnome blast was firmly in the spirit of a larger U.S. government initiative to make the best of the reality that it had inherited. The centerpiece was a magnanimous attempt to share nuclear technology so that countries such as Pakistan, India, and the Belgian Congo could maintain their own reactors and be partners in the revolution foreseen by William L. Laurence. The dream of atomic power had been slow to take root (the United States finally achieved critical mass in an experimental station in Idaho a full six years after Hiroshima; commercial use was still a decade away), but President Dwight Eisenhower took an active interest and was convinced that it ought to be shared with all friendly countries. They would get it anyway, he reasoned. Why not sow goodwill by simply giving it to them first? He foresaw "an atomic Marshall Plan" for the world in which the United States would donate about 44,000 pounds of

U-235 for distribution to qualified nations. This was, ideally, to be the start of a global uranium bank.

"It is not enough to take this weapon out of the hands of the soldiers," said Eisenhower in a landmark speech before the UN in 1954. "It must be put into the hands of those who will know how to strip its military casing and adapt it to the arts of peace."

This was the beginning of the controversial Atoms for Peace program, which planted reactors in such unlikely locations as Bangladesh, Algeria, Colombia, Jamaica, Ghana, Peru, Syria, Pakistan, Turkey, and the Belgian Congo, earning millions for Western power contractors. Eisenhower's program also led to the founding of the International Atomic Energy Agency, a thin ghost of the supercouncil first proposed by Bernard Baruch. The agency was supposed to promote the peaceful spread of uranium fission across the globe, with no mandate or funding to challenge nations that had other ideas. Opponents of the plan called it "Kilowatts for Hottentots," after the perjorative name for an African tribe, and complained that such sensitive equipment ought not to be spread so promiscuously.

But uranium cannot undergo fission in a reactor without producing a tiny residue of plutonium. This is an immutable law of physics. A "peaceful" nuclear reactor is no different in basic design from the complex at Hanford that manufactured the plutonium for the Nagasaki bomb. And herein lies one of the damnable paradoxes of uranium: The apparatus that spins a turbine also happens to be a munitions plant. One is a coefficient of the other; the mineral cannot escape its own unstable essence. "Mutually assured destruction" had been formulated with the Soviets in mind. Now here was a new idea: that somewhere, somehow, an unexpected actor with the means, the intellect, and the willpower could spirit away enough uranium to acquire the most formidable instrument in the world.

In fact, it was already happening.

"There's uranium in the desert. . . ."

The dreamy utterance floated like a prophecy across the dining room table in a New York City apartment in the fall of 1947.

The speaker was Ernst David Bergmann, a professor of chemistry who had been invited over for a supper of scrambled eggs by his friend Abraham Feinberg after the two had attended a Friday-night synagogue service together.

Feinberg was not happy to hear this prophecy spoken out loud in his dining room. He was then at the beginning of a career in the hosiery business; he headed the successful Hamilton Mills company and would one day acquire prestige clothing brands such as Catalina swimwear and Fruit of the Loom underwear. A lover of squash and cigars, he was also an outspoken advocate and fund-raiser for the rights of Jews to permanently settle in Palestine. Among his friends he counted President Harry S. Truman and also the farmer and journalist David Ben-Gurion, who would soon become the first prime minister of the newborn state of Israel. Feinberg was a cultured and discreet man, a graduate of law school who never practiced law, and he didn't want to hear this loose talk about uranium. It seemed rash. He told Bergmann to keep his mouth shut about it.

But Bergmann was no dreamer. He knew exactly what he was talking about. For the last twelve years, he had been quietly studying ways to build up a technological edge for the Jewish paramilitary groups in Palestine who had been agitating against the British for an independent state. Bergmann's position was not clandestine, but he was well placed as the scientific director of the Daniel Sieff Research Institute near Tel Aviv. He had trained many of the scientists who would make seminal contributions to the new Israeli military, including advances in rocketry, firearms, and surveillance equipment. Along with Feinberg, he enjoyed a personal friendship with Ben-Gurion. Yet the uranium in the desert was a topic too sensitive to be discussed in the open.

The "desert" Bergmann was talking about exploring was the Negev, a picturesque expanse of sandstone cliffs, dry riverbeds, and basalt mountains in the southern quarter of Palestine, a depopulated region that would soon be absorbed into the new state of Israel. The "uranium" referred to the ore in some of the northern Negev's phosphate deposits that Israel would soon be mining and using to stoke a secret reactor. The Israelis would eventually use the resulting plutonium to build up an arsenal whose total has been estimated at between two hundred and four hundred warheads. The number remains classified today, as does the very existence of the program.

Israel's journey toward nuclear armament was one that involved high-level deception, overseas financing, careful diplomacy, and, at one point, an elaborate charade of sea piracy on the Mediterranean, but it began with the most basic question of atomic potency for any nation: Where do you find the uranium?

In the case of Israel, it was virtually lying in the backyard. Uranium in the land of the Bible was on the minds of the leadership before Israel had even won its war of independence against a coalition of Arab foes in 1948. When the Negev was still largely controlled by the Egyptian army, a group of scientists was sent to locate and assess the deposits Bergmann had described. The head of the exploration team was a thirty-year-old chemist turned solider who confirmed that the Oren Valley indeed had a small source of low-grade uranium.

His team had been acting on a well-sourced tip. A Jewish paramilitary group called the Haganah—the forerunner of today's Israel Defense Forces—had learned that a British mining company had discovered some mysterious "black rocks" near a source of crude oil and phosphates. Bergmann figured he could extract at least five tons of uranium from the Negev each year and feed it directly into a heavy-water reactor.

This was exactly what David Ben-Gurion wanted to hear. The idea of an Israeli nuclear program enchanted him from the start of his tenure as prime minister in 1948. The new but small Mediterranean state was boxed in by a ring of hostile Arab nations that would need to win only a single decisive military victory to eliminate it. The Holocaust had proved that anti-Semitism could lead to death on an industrial scale, and Ben-Gurion felt that Israel had to possess the ultimate tool of defense—if not immediately, then at some point in the future. The state would have to become a coastal fortress with an awesome weapon of final resort. Though he was not a particularly religious man, there was a touch of the sacred in the way Ben-Gurion viewed atomic weapons. "This could be the last thing that could save us," he said once in a speech, referring vaguely to "science," but knowledgeable listeners heard the intellectual rumblings of what would become Israel's own Manhattan Project.

Ben-Gurion was also an enthusiastic promoter of the Negev, where he made his home, the same place where Abraham and Jacob were said to have tended flocks of sheep. In 587 B.C., the prophet Ezekiel had been commanded to go to the Negev and say on God's behalf: "Behold, I am about to kindle a fire in you, and it will consume every green tree in you, as well as every dry tree; the blazing flame will not be quenched and the whole surface from south to north will be burned by it. All flesh will see that I, the Lord, have kindled it; it shall not be quenched." An atomic reactor could "make the desert bloom," Ben-Gurion told weekend visitors, by desalinating water for agricultural use. And so much the better if the uranium could be drawn from the Negev, too. Development was initiated under the cover of a fertilizer company innocuously called Negev Phosphates Chemicals Co., and three small plants were built to separate the small bits of uranium out from the grayish sand.

Israel's real break came in 1957, when France found itself mired in a diplomatic crisis with Egypt over the Suez Canal and, in need of a regional ally, secretly agreed to provide technical and construction assistance for a reactor in the Negev south of the town of Dimona. Concrete footings were poured the following year, and false stories were spread that the Israeli government was building a "textile plant" or, alternatively, a "metallurgical lab." Most of what was built was built underground, in an attempt to hide its true nature from airplanes and spy satellites. Groves of palm and other trees were strategically planted around the facility to obscure the view of the containment dome. The twenty-five hundred workers on the site received their mail via South America from a phony post office. France also offered a prime supply of U_3O_8, otherwise known as yellowcake, the sickly colored powdered form in which uranium is typically barreled and transported. Ernst Bergmann supervised the reactor program from his post as chairman of the Israel Atomic Energy Commission, a body the stated purpose of which was peaceful research, but the real job of which was tending the subterranean facility at Dimona, where the heavy-water reactor first went critical in 1963. It was large enough to manufacture about fifteen pounds of plutonium per year, enough to build at least four bombs annually.

The financing had been almost entirely off the books. Much of it had come from a series of donors in the United States recruited by Abraham Feinberg, who had first heard the Negev uranium forecast over scrambled eggs and, according to the historian Avner Cohen, had had a 1958 meeting with Ben-Gurion about securing a stream of funds to be kept separate from the regular military budget. The contributions Feinberg helped raise, totaling an estimated $40 million, were euphemistically said to be dedicated to "special weapons programs." Fund-raising for Dimona was no ordinary charity work for Feinberg, who reportedly approached at least eighteen people in New York City about the project. This was at the highest level of religious and emotional significance, with the mysterious fission in the reactor core seen as a unique flame for the reborn nation.

"Ben-Gurion called the donors *makdishim*, or consecrators, and their contributions, *hakdasha*, consecration," the historian Michael Karpin has written. "Both of these Hebrew words derive from the word *kadosh*, sacred, which is also the root of the word Mikdash, or Temple—the holiest institution of Judaism. . . . And like the Temple, which was erected with the contributions of the children of Israel (Exodus 21:1), so too Israel's nuclear program would be built with contributions. In Ben-Gurion's eyes, the nuclear project was holy."

Discovery was inevitable. The CIA routinely examined photographs from U-2 spy planes worldwide and started noticing strange earth-moving patterns near Israel's "textile plant" almost as soon as construction began. The patterns closely resembled the work the French were doing on their own plutonium facility at Marcoule. When matched with reports that Norway had sold twenty tons of heavy water to Israel, this seemed to point toward an Israeli reactor that ran on natural uranium. Suspicions were relayed to President Eisenhower, who chose to ignore them. But after these reports hit the press in December 1960, Ben-Gurion was obliged to give an official explanation before the national legislature of Israel, the Knesset. He acknowledged that a twenty-five-megawatt nuclear plant had been built at Dimona "for peaceful purposes" and emphatically denied the existence of any plan to manufacture bombs. Two American visitors were given a brief tour and shown blueprints. But they were, in effect, shown nothing.

Before their visit, a particular bank of elevators was bricked over and the new wall concealed with fresh paint and plaster. These elevators were on the top floor of an unremarkable windowless concrete building near the reactor. Eighty feet beneath this building was exactly what the United States had feared: an automated chemical plant for the reprocessing of the plutonium recovered from the fuel rods in the reactor core. The material is highly radioactive and must be stored in sealed boxes full of argon gas; technicians insert their hands into prefitted gloves in order to shape the plutonium into precisely designed warhead components. This six-level facility, known as the Tunnel, was the birthplace of the Israeli bomb.

By the end of 1966, Israel had enough of the pinkish-orange metal to manufacture at least one usable nuclear weapon. Many more, perhaps as many as four hundred, have followed since. But there were no announcements or fanfare, and the bomb was never tested with a detonation. Analysts believe Israel has made multiple "cold tests," which involve simulating an explosion by testing all of the components separately.

The prime minister who succeeded Ben-Gurion, Levi Eshkol, formulated a policy of "nuclear ambiguity," or "opacity," that remains state doctrine today. In short, Israel will not publicly confirm or deny that it has the bomb, leaving the question permanently vague. Its leadership has promised on repeated occasions, "Israel will not be the first to introduce nuclear weapons in the Middle East."

But the word *introduce* was never defined. Does it mean "use in warfare"? Does it mean "threaten with nuclear weapons"? The word is a finely tuned cipher, containing a galaxy of possibilities. A dose of regional empathy was also involved, according to Michael Karpin. Arab countries were now handed a bulletproof excuse for *not* marching on Jerusalem— no country wanted to see its own capital turned to green glass—thus leaving intact the binding force that holds the Middle East together.

One thing is certain, though. The Israelis believe that "introduction" does not mean possession. The existence of the Tunnel has been called the worst-kept secret in the intelligence world. Two years after the first bomb was built, the CIA included the news in a National Intelligence Estimate and concluded that three more warheads were on the way. Any remaining doubts were shattered in 1986 when a technician named Mordechai Vanunu broke cover and gave a detailed description to the London *Sunday Times*, complete with illicit photographs of the six-level facility that lay behind the bricked-over elevators. Vanunu had become disenchanted with his nation's policy toward Palestinians and, after he was laid off from Dimona, took a backpacking trip around the world, ending up in Australia, where, as he wrote to an ex-girlfriend, he enjoyed the company of the locals because "they drink a lot of beer." He converted to Christianity while there and was eventually persuaded to go to the newspapers with his knowledge of the secret atomic bunker, in the belief that total disclosure would help bring a faster peace to the Middle East. He traveled to London to meet with the *Sunday Times*, where editors were skeptical and insisted on having Vanunu's assertions checked by a team of nuclear physicists before they would publish.

His assertions were serious and could have major diplomatic consequences. Israel had not only been building up a formidable atomic arsenal, it was also separating more plutonium than previously thought, as well as creating tritium and lithium 6, the reagents of massive thermonuclear devices. "These were weapons that could obliterate a major city," *Times* reporter Peter Hounam wrote in a later account. "They had no sensible battlefield application."

While he waited for the story to appear, the thirty-year-old Vanunu met a heavy-set blonde named Cindy, and the two struck up a fast friendship. Cindy said she worked for a cosmetics company in America and she listened with awe and admiration to his life story. They went to concerts and art galleries, and then she suggested they take a weekend getaway to Rome, where her sister had a flat where they could stay. That sounded like a good idea to Vanunu. When he walked into the flat, he was tackled by two men, injected with drugs, and smuggled onto a fast boat to Tel Aviv, where he learned that his new girlfriend, Cindy, had actually been a "honey trap," in the pay of the Mossad, the Israeli secret service.

His captors took him into a cell and thrust the front page of the October 5 *Sunday Times* at his face. It was headlined in bold type: REVEALED: THE SECRETS OF ISRAEL'S NUCLEAR ARSENAL. The paper and its team of experts had verified every detail of his spectacular story. "See the damage you have done!" they yelled. Vanunu was sentenced to eighteen years in prison for treason.

Despite overwhelming evidence of the Tunnel's existence, Israel maintained ambiguity. The story was neither confirmed nor denied. "I hope there won't be any more bother with this matter," Prime Minister Yitzhak Shamir told national radio after Vanunu was in custody. The arsenal was a forbidden topic, a black box, a secret wedged into the tunnel. An inscrutable silence, it was thought, had deterred Arab neighbors from clamoring for their own bomb in the name of regional balance.

This philosophy was not perfect, though, and one shortcoming had been exposed in 1968, when it created a major diplomatic problem. After years of negotiation and compromise at the UN, all the sovereign nations of the world, from China to Togo, were given the option of signing a global treaty regarding nuclear weapons. Aside from the Pax Romana and the UN itself, this treaty was arguably the first successful effort in the history of the earth that sought to extend a covenant to all known civilization; not just military superpowers, but every tiny principality in every distant ocean.

The Nuclear Nonproliferation Treaty, known as the NPT, was the essence of simplicity. The five nations with acknowledged atomic weapons—the United States, the USSR, China, Great Britain, and France*—would agree not to give away the weapons technology or the uranium. Those 180 nations who were the atomic "have-nots" would agree not to build weapons or manufacture highly enriched uranium, and to open their existing reactors for spot checks by experts from the International Atomic Energy Agency. The inspectors would be looking for evidence that plutonium was being carried off—in inspector lingo, "diverted"—for use in a bomb. This was exactly what had been happening in the Negev for the better part of five years.

---

* These nations also happen to be permanent members of the UN Security Council. Nukes have their privileges.

The NPT was elegant, but it was also crude. It was the logic of school-yard bullies, all with rocks in their fists and none wanting to be struck. To attack would invite destruction; a selfish peace would therefore be necessary. The treaty incorporated a cold military reality that had been plain to every serious policymaker since Hiroshima, and what had been neatly stated by the declaration of the Harvard and MIT scientists in 1945: that the blueprints were public and the uranium was plentiful, and international cooperation was the only thing that could prevent a border squabble from becoming an inferno. The "club of five" nuclear elite was not required to disarm and would thus be exempt from the indignity of reactor inspection. What would be the point of hiding uranium enrichment if you already have an arsenal?

A distasteful reality was therefore enshrined: Five nations on the globe would have the power evermore, and all the rest would have to take a vow of abstinence. The NPT, in the words of one analyst, was like a man with a cigarette dangling from his lips telling everyone else to stop smoking. A provision obligated the club of five to work toward "complete disarmament under strict and effective international control," but this blandishment was so vague as to be meaningless. A bipolar world was also guaranteed through this treaty: Uncommitted countries could pick one of two nuclear umbrellas to hide underneath in crisis, that of either the United States or the USSR. Choices had never been so narrow in a world without uranium. "Given the perverse set of values mankind has inherited from its violent past," one physicist noted, "a nation with the power of annihilation is presumably more important than one without it."

In spite of these acknowledged hypocrisies and shortcomings, the treaty picked up fifty-nine signatures the year it was presented and, after ponderous diplomacy, would eventually be signed by almost every country on earth except Israel, for whom the treaty was a serious conundrum. Refusing to go along would be a tacit admission that Dimona indeed concealed a sophisticated plutonium factory. But signing would mean an embarrassing disarmament and the end of the hard-won nuclear program. This would deny Israel the hole card that military strategists called the Samson Option, after the muscular antihero in the biblical book of Judges who pulled down the pillars of a Philistine temple to kill his enemies (and himself) rather than face death by torture. It was a Levantine version of mutually assured destruction: If faced with another certain genocide, Israel could use the ultimate weapon to vanquish its invaders, even at massive cost to itself. The secular Armageddon foreseen in 1940s America was viewed in the hard light of realpolitik in Israel, home of the original Megiddo. "The memory that no country was prepared to help when Hitler murdered six million Jews makes Israelis doubt that any country would come to their aid if they were being pushed into the sea," noted one analyst.

Israel ultimately refused to sign the NPT and thus remains one of only four nations outside of it (the others are North Korea, India, and Pakistan). A critical decision had been made: The Jewish state would remain the sixth member of the "nuclear club," albeit shrouded in fog. The Dimona reactor's megawatt capacity was eventually expanded to nearly six times its original French design. But it was also becoming starved for more uranium. The Negev mines were not producing enough, and the French, wary of international condemnation, had withdrawn their support after Charles de Gaulle secretly ended what he called "irregular dealings" with Tel Aviv.

The Israelis could hardly buy their uranium on the open market. Another source would have to be secured. This was the genesis of a fake pirate operation in the Mediterranean that would eventually be known as the Plumbat Affair.

Plans were elaborate. Agents from the Mossad set up a fictitious company based in Liberia and purchased a tramp ocean freighter they rechristened the *Scheersberg A*. They then enlisted a friendly mid-level official at a German petrochemical company who arranged for the purchase of $3.7 million worth of yellowcake from Union Minière. The uranium had apparently been mined from Shinkolobwe several years prior, and the Belgian company was trying to rid itself of the final inventory. A contract was arranged to have it processed by a paint company in Italy that had never before handled such a large quantity.

In November 1968, the *Scheersberg A* was sent to the wharves at Antwerp to pick up the uranium. Two hundred tons of it were loaded into the ship, packed into barrels stamped with the misleading legend

PLUMBAT, which is a harmless lead product. The entire crew of Spanish-speaking sailors was fired and a new crew, composed of Mossad-selected workers with forged passports, took their place. The ship eased out of port at sunrise on November 17, supposedly heading for Genoa and the paint company. But approximately seven days after leaving port, in the middle of the Mediterranean somewhere east of Crete, the ship made a nighttime rendezvous with an Israeli freighter.

"As two Israeli gunboats hovered near the freighters, the barrels were transferred in total darkness," reported *Time* some months later. "Except for an occasional Hebrew command, no one spoke." The freighter bearing the yellowcake sped off eastward toward Haifa, and eventually the Tunnel.

The *Scheersberg A*, meanwhile, docked in Turkey eight days later with no cargo. Several pages had been ripped out of its logbook. The Italian paint company was told to cancel the $12,000 processing contract because the uranium had disappeared. No further explanation was given, and the Italians were left to assume that an act of piracy or a hijacking had occurred.

The true story might have never come out if it weren't for an alleged act of desperation on the part of a Mossad agent named Dan Ert, who was arrested five years later in Norway on suspicion of having helped assassinate one of the Palestinian terrorists who had killed eleven Israeli athletes at the 1972 Olympic Games in Munich. Confined to his cell, and apparently panicking, Ert confessed his identity as an intelligence operative and, to prove it, related the story of the secret uranium transfer from the *Scheersberg A*. His story gained credibility after investigators discovered that he had been listed as the president of the Biscayne Trader's Shipping Corporation, the shadowy outfit in Liberia that owned the ship. Ert was convicted of participating in a murder; he served seven months in a Norwegian jail.

The Plumbat Affair leaked out in 1977 when a former U.S. Senate attorney named Paul Leventhal spoke about it at a disarmament conference in Austria. The stolen uranium shipment, he said later, was enough to run a reactor such as Dimona for up to a decade and could yield plutonium for up to thirty atomic weapons. He added that any country that wanted uranium could probably obtain it because "safeguards are so weak, incomplete, secretive, and slow." The *Los Angeles Times* quoted an expert who characterized the midnight sea transfer not as a hijacking, but as a "laundering" of illicit yellowcake.

Israeli officials reacted at first with silence and then professed total ignorance about what happened aboard the *Scheersberg A*. "We deny all aspects of the story which relate to Israel," said a spokesman, days after the news broke. The European Atomic Energy Commission pledged to tighten the rules on the sale and transfer of uranium and other sensitive materials. But no substantial changes were made. Nuclear ambiguity carried the day.

Yet Israel was not the only country that hoarded uranium in the name of staving off the apocalypse.

Pakistan was founded as a religious homeland at almost exactly the same time as Israel, and also in the dust of British retreat from Empire. Its borders were made by the slash of a pen in 1947, the handiwork of a beleaguered colonial official sent from London to separate Hindu from Muslim as quickly as possible before the British flag was lowered for good.

The nation's name, too, is bureaucratic artifice; an acronym coined in the 1930s by university students at Cambridge. It happens to mean "pure land" in Persian, but the letters were originally supposed to represent a confederation of Muslim regions struggling for independence from their pantheistic neighbors: *P* is for Punjab, *A* for the Afghan mountains, and *K* for the gorgeous mountain province of Kashmir.

This last place would become the scene of two bloody and humiliating wars with India, which added to the insult by stating its interest in building an atomic bomb in the mid-1960s. A rising political star in Pakistan, Zulfikar Ali Bhutto, responded to India's threat by telling a newspaper that his nation "will eat grass or leaves—even go hungry—but we will get one of our own." Bhutto was already famous for this kind of rhetoric, a blend of the homespun and the apocalyptic, but he had a genuine fixation on the technological edge that India had gained. For Bhutto, there was no better yardstick of progress than an atomic weapon, and so when he took over as president in 1971, it was foreordained that a nuclear program would be under way.

The stakes climbed after India test-exploded a nuclear device named Smiling Buddha in an underground shaft in 1974. This test was made—pointedly—several dozen miles from the Pakistan border, and it became a watershed moment in the lives of many Pakistanis who shared their leader's obsession with maintaining a balance of military power with the Hindu superpower to the east.

Among those Pakistanis was a thirty-six-year-old technician named Abdul Qadeer Khan, who was then living in a rural town in Holland, a short drive from Amsterdam. He was moved to write a letter to Bhutto offering his technical services. This was no vain act of patriotism: Khan was in a genuine position to help. After earning a doctorate in metallurgical engineering in Belgium, Khan had landed a job at a subcontractor for Urenco, Europe's only facility that enriched uranium for nuclear power plants. Though he had only a mid-level security clearance, people liked the outgoing and charming Khan, and he was permitted access to the sensitive areas of his own plant, and even Urenco's. He could therefore offer what Pakistan most wanted: an easy path to the bomb.

Khan flew home for a Christmas break in 1974 and managed to get a personal audience with the prime minister. There he made a case for uranium that changed the course of history. Bhutto had been committed to the same route that Israel had taken: diverting plutonium from a nuclear reactor. In this case, it was Pakistan's new 137-megawatt reactor at Karachi. But after listening to Khan's briefing, Bhutto became convinced that it would be easier and safer to pursue a uranium solution. The material was safer to handle, for one thing. And it could also be concealed more easily from Western spies.

Bhutto authorized Project 706 in which army units were dispatched to the Siwalik Hills to secure deposits of low-grade uranium ore, and ground was cleared for an enrichment plant in the desert near Rawalpindi. But as the Manhattan Project scientists had learned in the United States thirty years before, enrichment is a job of titanic industrial proportions. The gas centrifuge method sought by Pakistan would require a pipeline of thousands of centrifuges through which a thick cloud of gaseous uranium would be forced.

From the outside a centrifuge looks a bit like an elongated scuba tank. Inside is a steel or aluminum rotor that spins at fifteen hundred revolutions a minute, driving a slow but effective wedge between U-238 and its lighter (and more volatile) cousin U-235, a bit like a churn that separates butter from milk. The working parts of a gas centrifuge—the rotors, molecular pumps, suspension bearings, and baffles—must have precise metallic alloys and exact designs because the centrifuge turns at a velocity of near the speed of sound and must spin continuously for years. An imbalance as tiny as a microgram can cause the whole thing to whiz off its bearings and make a spectacular crash. The International Atomic Energy Agency restricts the export of centrifuge parts to guard against the possibility that someone might try a feat of reverse engineering. But just as Israel found a way to move a few hundred barrels of uranium into the Mediterranean, where they could be quietly transferred, A. Q. Khan found a vulnerability that could be exploited.

Khan began to steal materials and blueprints from his employer and funnel them to Pakistan via diplomatic pouch from the embassy in Amsterdam. He also had a close friendship with Frits Veerman, a coworker and photographer, whom he asked, with increasing frequency, to take photographs of highly sensitive equipment. Veerman assumed that Khan's project was authorized and, in any case, he had no reason then to question his friend's motives. The two men enjoyed ogling pretty girls on Amsterdam streets as much as the newest enrichment gadget at the lab. Veerman was an occasional dinner guest at Khan's house—the menu usually involved barbecued chicken and rice cooked by Khan's South African wife, Hendrina—but eventually Veerman began to suspect something was amiss with his friend.

"Top-secret centrifuge drawings were lying around in Abdul's house," he recalled to an interviewer years later. "They were only supposed to be used at the plant and stored in vaults there afterwards. That was my biggest worry. What was he doing with those drawings? All the little pieces of the jigsaw put together made me reach the conclusion that Abdul was spying."

Veerman tried to sound the alarm. First he made an anonymous call to the authorities from a pay phone. When that failed, he went to the manager of the laboratory and was reprimanded for making wild allegations. Khan was nevertheless transferred to a management job, where he had no more direct contact with sensitive equipment. But his dirty work was already finished: He had the technical specifications for a factory that would allow any nation to fashion an atomic bomb out of raw uranium.

At Christmas 1975, one year after his private meeting with Bhutto, he flew back to Pakistan on a routine vacation and did not return. Hendrina wrote a letter to a friend explaining that her husband had gotten sick, delaying their return. He mailed in his resignation two months later. Bigger things were now in store. Khan founded Engineering Research Laboratories, a company with a remit directly from the prime minister to build an enrichment plant inside an unruly agrarian country where most people lived in poverty and running water was a luxury. Khan himself later marveled that a nation that could not even manufacture sewing needles or bicycles was attempting to reproduce the most futuristic technology on the planet.

A good set of drawings had been cribbed from the Dutch, but Khan didn't yet have all the machinery. This problem, too, called for deceptive tactics. Khan and some associates took advantage of weak export laws to buy, from Swiss and German companies, parts that could be used for innocent purposes as well as for uranium enrichment—high-strength aluminum tubes, for instance. There were guidelines from the IAEA for the exports of these "dual use" technologies, but governments were spotty about enforcing them, and Khan found them easy to circumvent. He would often hide a critical purchase within a long shopping list of innocuous items, for example, or have the equipment shipped to phony buyers in different countries to disguise their true destination. Some metallic components were acquired as "sales samples" and then taken to his lab in the town of Kahuta for study and reproduction.

Khan was also willing to pay well above the market price, which should have raised suspicions but instead made him a favored customer of European firms eager for overseas accounts. Tracing Khan's trail many years later, a team from the London-based International Institute for Strategic Studies (IISS) concluded that Western greed played a major role in Pakistan's fortunes, as did hubris. "Many industrialists reasoned that 'if we do not do it, others will,' and deliberately violated the law. A willful naivety and arrogant skepticism about Pakistan's ability to put sophisticated machinery to military use also played a role."

The centrifuges turned, and the uranium poured out. By 1984, Khan claimed that Pakistan was ready to set off an atomic bomb with a notice of only one week. This may have been only a slight exaggeration. Less than two years later, the CIA reported that Pakistan was "two screwdriver turns" away from arming a weapon, which had likely been cold-tested at Kahuta. When India's military held a provocative exercise near the border, Khan felt confident enough to brag to a reporter, "They told us Pakistan could never produce the bomb, and they doubted my capabilities, but they now know we have done it."

He then made a statement to a newspaper that put him in a role as the nuclear mouthpiece for the government. "Nobody can undo Pakistan or take us for granted," he said. "We are here to stay, and let it be clear that we shall use the bomb if our existence is threatened." Khan tried to complain he had been misquoted, but it happened to have been the truth. Earlier exposés about the program in the German media had only helped him, as new European firms starting calling on him with offers to sell dual-use products. "In the true sense of the word, they begged us to purchase their goods," he said later. "And for the first time, the truth of the saying 'They will sell their mothers for money' dawned on me."

Pakistan emerged from the nuclear shadows in full in 1998, when it tested five weapons inside a mountain bore and left clear seismic evidence of its new powers. This was the first time an Islamic government—and a poor one at that—had direct control of the tools of doomsday, and it created an enormous burst of national pride and confidence. Pakistan now had the ultimate negotiating tool in world affairs. Khan was hailed as a genius and was venerated in a cultish way that perhaps only a Manhattan Project scientist might have recognized: celebrated for birthing a tool of death. Schools and cricket teams were named for him; boys wanted to *be* him when they grew up. He could not go to a restaurant without somebody buying him dinner. Outstretched hands greeted him wherever he went. Khan told an interviewer: "If I escort my wife to the plane when she's flying somewhere, the crew will take notice of who she is and she will receive VIP treatment from the moment she steps on the plane. As for me, I can't even stop by the roadside at a small hut to drink chai without someone paying for me. People go out of their way to show their love and respect for me." For a beleaguered nation, resentful of Western technological suzerainty, Khan and his uranium bombs seemed to provide an avenue of hope: a sense that Pakistan, too, belonged in an elite group along with its hated neighbor to the east with which it shared a 1,800-mile border—about 200 miles shorter than the border between the United States and Mexico—and only a single connection with a paved road.

Paid handsomely by a grateful government, Khan built a lakeshore mansion outside Rawalpindi, not far from the uranium fields. He gave cash to charitable foundations, funded mosques, was photographed feeding his pet monkeys (a daily diversion), and continued his work in the national labs, which had been renamed Khan Research Laboratories in his honor. In the romantic city of Timbuktu in Mali, he built a luxury hotel and named it after his wife. A Pakistani air force C-130 cargo plane was used to ship exotic wooden furniture from Islamabad to the hotel lobby. Dissidents in Pakistan claimed that the Hendrina Khan hotel was supposed to have been a base for desert uranium prospecting; this remains unproved.

"In his middle age he had become a fleshy, banquet-fed man, unused to criticism, and outrageously self-satisfied," wrote the journalist William Langewiesche. "Accompanied by his security detail, he went around Pakistan accepting awards and words of praise, passing out pictures of himself, and holding forth on diverse subjects—science, education, health, history, world politics, poetry, and (his favorite) the magnitude of his achievements." He bribed local journalists to write puff pieces on him and reportedly endowed several charities on the condition that they shower him with awards.

Khan might have remained this way, a benevolent combination of nuclear avatar and National Uncle, but something had been driving him to keep pushing his mandate. Though he had been convicted in absentia of "attempted espionage" back in Holland, he always denied stealing any blueprints and resented the implication that his own talents lay in burglary rather than honest science. This was too much for his ego to bear. "Khan felt his capabilities had been insulted," said the IISS experts. "He may also have felt a genuine sense of injustice and a victim of hypocrisy given the high number of Western industrialists who were more than ready to do business with him."

There was also the inherent hypocrisy of the nonproliferation treaty to consider. In the eyes of Khan and most Pakistanis, the Americans and the other first-class powers were ready to shake their fingers at any nation that wanted to develop a nuclear security blanket, yet they maintained huge arsenals of their own. This was a special affront to the Muslim nations of the world, which had been the intellectual and military masters of the globe in the ninth century but now saw themselves under the nuclear boot heel of the West.

Though he prayed five times a day and made public thanks to Allah for his good fortune, Khan was no fundamentalist. He also displayed no interest in sharing his technology with terrorists. But he believed in Pakistan's right to enrich its own uranium and command its own respect, just as other beaten-down nations had the right to do.

Whatever his motivations—narcissism, profit, altruism, revenge—Khan and his employees had been making undercover sales of Pakistan's uranium and nuclear goods to a variety of pariah countries, including Iran, Libya, and North Korea, for more than fifteen years before the network was finally undone by the seizure of a ship in a Mediterranean port.

The band of dealers now known as the A. Q. Khan Network or, more flippantly, as the Nuclear Wal-Mart, apparently made their maiden sales call in 1987 with an offer to sell a sample of centrifuge parts and plant drawings to Iran. It was basically the same atomic starter kit that Khan had smuggled out of Europe in the previous decade. Iran was then locked in a stalemated border war with Saddam Hussein's Iraq and was looking for a way to show an advantage. Iran bought a portion of this shopping list for $3 million and went to work.

That was only the beginning. Khan began to make phone calls to his old contacts at the European suppliers and shell companies he had used to build Pakistan's enrichment plant—only now he was interested in exporting, not importing. The beneficiaries would be any nation that wanted to make a deal with him. As it had been before, his favorite tool was uranium. Plutonium was never an item that Khan wanted to trade, as it required care in handling and could kill its courier. It was also beyond his level of technical understanding.

A top customer was North Korea, which traded some of its ballistic No-dong missiles for, as Khan later confessed, "old and discarded centrifuge and enrichment machines together with sets of drawings, sketches, technical data" and uranium hexafluoride gas. These castoffs were not enough to create the ball of uranium required for critical mass, but the paranoid Kim Il Sung—"Dear Leader" to his people—apparently tried to reverse engineer an enrichment plant of his own. Khan personally traveled to at least eighteen different countries in an apparent attempt to drum up new Third World customers for his nuclear goods, and visits to Nigeria and Niger may have been part of a plan to secure an underground stream of raw uranium ore.

The network operated under the cover of the Khan Research Laboratories, which boasted its own schools, hospitals, and a cricket team and blatantly advertised its services with glossy brochures and a sales video in which Khan's own voice is heard. "Together we can really work wonders," he says. The degree to which the Pakistani government knew of his activities is still a matter of conjecture, but it is beyond question that the head of the nation's nuclear program was given wide latitude to do what he pleased, and without monitoring or challenge. "Khan had a complete blank check," a top military aide later told a reporter. "He could do anything. He could go anywhere. He could buy anything at any price." If A. Q. Khan felt any guilt, he never showed it. "Who the hell is going to use nuclear weapons?" he reportedly said. "I see them as peace guarantors."

Khan chose a logical place to consummate most of his deals: Dubai, known as "Manhattan in the desert," home to a row of luxury high-rises, shopping malls, gold markets, and a busy airport ringed with free-trade zones. The richest city in the United Arab Emirates, formerly a sleepy outpost for pearl divers, began its climb to international prominence in the 1970s when the ruling al-Maktoum family made the decision to reinvest the emirate's modest oil royalties in the infrastructure, particularly the $3 billion shipping terminal. A favorable tax climate and a strategic location between China and the West added to its allure as a laissez-faire trading and financial center. A quarter of the world's construction cranes are now in Dubai, as is the world's tallest skyscraper, three artificial islands in the shape of palm trees, battalions of luxury cars on the freshly paved roads, an indoor snow-skiing mountain, an underwater hotel, and a massive amusement park called Dubailand, which, when completed, will be larger in area than the principality of Monaco. The U.S. Navy uses Dubai as a maintenance center; it is its most frequently visited port outside the continental United States.

But an atmosphere of perfidy still lingers in the desert city-state. The steel-and-glass downtown is bisected by an anemic waterway nicknamed "Smuggler's Creek," a reference to the place's history as a base for the wooden dhows that shipped liquor, soap, appliances, and other goods around the Arab world, sometimes behind false bulkheads in order to dodge taxes. The river of money flowing through Dubai conceals a significant trickle of illicit commerce. Mafia figures from Russia and India have been known to park their earnings in some of the splendorous real-estate developments on the shores of the Persian Gulf. Most of the funds for the September 11 hijack plot were wired from Dubai banks. Blood diamonds, stolen property, and sensitive military technology pass through the city under the names of dummy corporations. The local police lack the budget or the mandate to do much about these things.

John Cassara, a former U.S. Treasury official, wrote that he once tried to explain to a Dubai broker the necessity of making accurate invoices to show the customs agents. "With complete and obviously sincere innocence, he told me, 'Mr. John, money laundering? But that's what we *do.*' "

This was the perfect place for A. Q. Khan, who rented a penthouse apartment for himself and began to make deals.

As many as thirty shell companies may have stored or shipped his atomic material at one time or another. Skyscraper hotel rooms were occasionally used as meeting spots. Khan's most trusted subordinate, Buhary Syed Ali Tahir, acted as the network's Dubai branch manager. His job as the director of a family-run computer importer named SMB Group gave him the necessary cover.

Through Tahir's office, the network began doing uranium-related business with Libya's eccentric Mu'ammar al-Gadhafi, who had always had imperial designs on North Africa. In 1971, he had launched an invasion of Chad in a bid to gain control of a desert said to be rich with uranium. The adventure turned the dirt-poor Chad into an even more desiccated wreck and ended in an expensive defeat for Gadhafi. He later managed to purchase 2,263 tons of yellowcake from Niger, but he lacked any mechanism for improving it to weapons grade. Khan agreed to sell him centrifuges for use at a plant on the outskirts of Tripoli that could have produced up to ten bombs per year. The deal also involved the transfer of design papers (which came wrapped in a plastic bag from a dry cleaner), containers of uranium hexafluoride gas, and a steel device to feed the heavy gas to and remove it from a centrifuge. This last gadget was allegedly manufactured inside an ordinary-looking industrial shed in a suburb of Johannesburg, South Africa, by a company called TradeFin and under the supervision of a Swiss citizen named Daniel Gieges. The device was two stories tall. Employees nicknamed it "the Beast."

Libya might have succeeded at threatening the world with its own bomb by 2008 if Western intelligence agencies hadn't been paying attention. The CIA had begun assembling a file on A. Q. Khan almost from the day he left Holland, and reportedly told Dutch investigators *not* to arrest him in the mid-1970s because it wanted to keep him under surveillance. The case for arresting him was strengthened after IAEA inspectors made a visit to Iran's enrichment plant and found Pakistani designs. When U.S. officials learned in October 2003 that crates of centrifuges marked as "used machine parts" were on a freighter called the BBC *China* and heading for Libya, they persuaded the ship's owners to divert it to the port of Taranto in Italy, where it could be boarded and inspected. The paper trail pointed toward Tahir's shady computer company in Dubai. This embarrassment caused Gadhafi to make a spectacle of renouncing his weapons program ("This was like killing my own baby," lamented his nuclear chief) and mending a toxic relationship with the United States, one that had lasted for more than two decades.

Meanwhile, in Pakistan, President Pervez Musharraf was under pressure to do something about his rogue weapons scientist. Khan had already been fired as head of the research laboratories, and Musharraf had no choice but to place him under permanent house arrest. Knowing that he was handling an icon, though, he offered Khan a conditional pardon and refused to allow foreign intelligence agents to interrogate him about the activities of the network. Investigators were left to chase middle players. After Gieges was arrested, he complained he was a victim of atomic hypocrisy—much the same argument that Khan had made.

"What qualifies the Americans to have in excess of ten thousand nuclear explosive devices just waiting for someone to push the button?" Gieges wondered. "What qualifies the Americans to have this and not others?"

None of the old defiance was evident in a speech Khan made on national television following his arrest. It was a statement he had almost certainly been forced to make, and it was delivered in English, not Urdu, suggesting that the message was aimed at the West and not for a home audience. He said he was sorry, but his regret was not about the emporiums he had run from his laboratories. He seemed more concerned that the shield of uranium he had erected for his country had been sullied.

"It is with the deepest sense of sorrow, anguish, and regret that I have chosen to appear before you in order to atone for some of the anguish and pain that has been suffered by the people of Pakistan . . . ," he began. "I am aware of the vital criticality of Pakistan's nuclear program to our national security and the national pride and emotions which it generates in your hearts."

There is an institution that is supposed to break up a national love affair with uranium, and it can be found on the mealy plains south of Vienna, among wheat fields and birch trees. A rectangle of barbed wire surrounds a bland industrial park south of the village of Seibersdorf, and to get in, you have to pass through an airlocklike chamber in the guard station that fronts the highway. Inside the perimeter is an unremarkable two-story building of red bricks and tin trim. It was built in the late 1970s and would look at home atop a Wisconsin paper mill. There is no sign outside.

This is the home of the Safeguards Analytical Laboratory of the International Atomic Energy Agency, and its employees are responsible for making sure that every country that signed the nonproliferation treaty is not squirreling away uranium in a warehouse or trying to siphon off plutonium. The work of these nuclear inspectors can move global events, even when they find nothing of value. Saddam Hussein's initial refusal to cooperate with this laboratory helped lay the foundations for the U.S. war with Iraq, although he had given up his pursuit of uranium in the early 1990s.

I went there in January 2007 to meet the chief of the facility, a trim and earnest man in his mid-thirties named Christian Schmitzer, who took me through another guard checkpoint and into his office to explain exactly how an atomic crime can be detected.

"This whole regime hinges on mistrust," he told me. "When we go into a reactor, we're saying, 'We don't trust you.' Sometimes this causes tensions."

Inspectors travel on a rotational basis to places such as Argentina, Japan, and Sweden. They are led into the plant by (usually) friendly executives who offer them tea or coffee and then stand back while the inspectors put on latex gloves and produce a set of cotton balls as big as oranges. These are wiped across the pipe joints and manifolds, where traces of uranium dust and other process materials are prone to leak. The inspectors also make swipes inside the employee locker room, which yields the best samples of any room in the facility. People always shake their jumpsuits and lab coats when they change clothes, and the dust is everywhere.

The cotton balls are placed in plastic bags. Fuel pellets are counted. The inventory books are scrutinized, as are the suppliers' delivery records, and the two must match exactly. The air is checked with a gamma-ray spectrometer and, occasionally, a small patch of dirt from the plant's ground is dug up and bagged. A water sample might also be taken. If the facility happens to be an enrichment plant, the inspectors will go tap on the metal kegs of uranium hexafluoride with a hammer. Empty kegs make a distinctive gonging sound, and the approximate gas level must be matched with what the plant has reported. There are handshakes and good-byes. The inspectors mail off their "hot swipes" and dirt samples via an international parcel service.

Once back in the lab in Austria, the materials are assigned a random code so that nobody except the directors knows which nation they came from. They are taken into the "clean room," where everyone must wear fabric caps and overshoes. The cotton is scrutinized for any trace of U-235 with electronic microscopes and spectrometers that can resolve down to the femtogram, which is a unit of weight a millionth of a millionth lighter than a paper clip.

Enriched uranium has a distinct isotopic signature. The technicians in the lab know what uranium is supposed to look like in a power plant. If anybody was trying to turn it into weapons, the signs would be all over the place, like a murdered corpse that won't stop bleeding. "Even if they were trying to clean it up, they never could," said Schmitzer.

There are three basic ways in which a nation (or a rogue faction within) might try to pilfer some fissile material to make a bomb. The first is to simply steal it outright. This is apparently what happened with the uranium rods in the Democratic Republic of the Congo that later turned up in a Mafia deal in Italy. Known as a gross defect, this is the easiest discrepancy to see. The second way involves siphoning off a large portion of material, such as a few cubic meters of uranium hexafluoride, while leaving the rest in place. This is known as a partial defect. The last method is what an embezzler might call the salami technique. You shave a tiny bit of uranium off a large quantity of deliveries, like ultrathin slices from a salami, over time and hope nobody notices. This is known as a bias defect and would still be difficult to pull off. Both the suppliers and the facility would have to rig the paperwork.

A major weakness in the Safeguards lab's remit was exposed in 1992 after Saddam Hussein refused to allow inspectors into his suspected nuclear plants. Prior to that point, the primary job of inspectors was to check declared inventories of uranium against the actual levels, taking samples only from what the host nation claimed was there. But Iraq had never declared its program. The expected paperwork, therefore, didn't exist. Inspectors expanded their brief to include environmental sampling and forensics to determine the extent of programs that were officially secret, and to develop protocols for entering places where they were not welcome. It would not be exaggerating to say that what happens in this two-story brick building can trigger a war.

"I don't know of any lab that analyzes substances with more scrutiny than we do," said Schmitzer. "We have to consider what we do very hard. We must not screw up."

He repeated this, to make sure I understood.

"We must not screw up. There will be major political implications."

His police work is only as good as his access. Schmitzer's agency is powerless to learn anything about the bomb-making apparatuses in Israel, India, and Pakistan because those nations have never signed the nonproliferation treaty and have no reason to admit IAEA inspectors.

A fourth nation is a wild card. Having done brisk business with A. Q. Khan, North Korea withdrew from the treaty in 2003, citing "grave encroachment upon our country's sovereignty and the dignity of the nation." High-level pressure is necessary in cases like this. All the inspectors can do from Austria is to stay vigilant of those countries still inside the framework that might be buying black-market uranium or centrifuges.

Apocalypse—or a small-bore version of it—is also not a concept that belongs exclusively to governments. The fears of Western intelligence operatives also focus on terrorist groups or religious zealots who may find a dealer more willing than A. Q. Khan to sell them tools of fissile destruction. They would, of course, be immune from the prying eyes of the IAEA. And some movements might come from unexpected places, the kind of thing that no rational model can predict.

One example of this was a meditation group started in a Tokyo apartment in 1984 by a former yoga instructor. He called himself Shoko Asahara and taught a blend of traditional Buddhist enlightenment, mingled with the millennial imagery in the book of Revelation. He was a charismatic man, and the circle soon became trendy among young university graduates seeking a spiritual side to their lives. Asahara promised an antidote to what he called the "emptiness" of modern life. The group called itself Aum Shinrikyo—*aum* for the traditional meditative chant, the Hindi word for "universe," and *shinrikyo*, which translates as the "truth of the universe."

Asahara recruited younger members with his own line of graphic novels, rendered in the popular Japanese styles of manga and anime. They were quirky and fun, usually depicting spaceships and futurist gadgets operated by heroic characters fighting dark conspiracies and looking for the secret of the universe. William L. Laurence of the *New York Times*, who had wanted to fly an airplane to Mars, might have been intrigued with some of the more benign trappings of this group, particularly the emphasis on science and hidden meanings of life.

The teachings got weirder. One of the comic books featured a character who walked into a room and said, "My guru, the god Shiva suddenly said to me, 'Now is the time described in the book of Revelation; receive the message and start Aum's salvation work."

The exact nature of this task soon became clear. Asahara believed that he was destined to trigger a mighty global confrontation between good and evil, after which history would come to a close. Japan itself must suffer "many Hiroshimas," he taught, which would cause the superpowers to destroy the entire world. Killing innocent people was not a sin, but a blessing, because the victims would be released from the earthly cycle and bring more blessings to those who had murdered them. (It was a theology not unlike that of the Thugee cult of nineteenth-century India, whose members befriended lone travelers and ritually strangled them with a yellow sash.)

Asahara's chosen tool was uranium. He sent a team of senior associates to Western Australia to buy a piece of isolated land where the mineral was known to occur naturally. They paid nearly half a million dollars for a sheep ranch near the settlement of Banjawarn, where they said they were conducting "experiments for the benefit of mankind." The seed money had been raised by donations from new members (as well as the sale of Asahara's blood to the faithful—he charged $12,000 per thimbleful). The cult took out mining licenses in the names of two shell companies, Clarity Investments and Mahapoysa Australia, and began to dig. An Australian geologist was consulted about the feasibility of bypassing national export laws and quietly transporting the ore by ship to Japan. Computer-savvy members also hacked into government databases to steal information on the enrichment process. By this time, the cult had become popular enough to attract upward of ten thousand members.

The doomsday plans hit a snag in 1993 when members tried to carry generators, picks, gas masks, and hydrochloric acid (in bottles marked HAND SOAP) on board a commercial airline flight into Perth. Authorities charged two men with carrying dangerous goods on an airplane, but released them. The sheep ranch yielded a small amount of uranium, which the cult had planned to enrich with the help of nuclear scientists recruited from Russia. These experts were later described as "second rate" by investigators, but it hardly mattered because the cult had managed to mine only a small amount of uranium ore. There was not nearly enough for a weapon.

Asahara was growing impatient and depressed. He had predicted the end of the world would commence in 1995, and the carnage had to begin before then or he would risk embarrassment. He told his members to test sarin gas on some of the sheep on the Australian ranch. Authorities later found a field teeming with animal bones.

Sarin is an odorless nerve agent that hangs thick in the air and causes violent spasms and death, a reaction not unlike that of a cockroach upon being sprayed by pesticide. It was first formulated, but not used, by Nazi scientists during World War II. A single drop is powerful enough to kill a healthy adult. On March 20, 1995, five teams of two men each descended into the subway stations at the heart of Tokyo and boarded separate trains. They carried bags full of newspapers soaked with about a liter of homemade sarin. At approximately 8 a.m., they punctured the bags with the sharpened tips of their umbrellas and ran away. Twelve people died and more than a thousand were hospitalized.

Asahara issued a press release denying responsibility for the attack— "We are Buddhists! We do not kill living beings"—but when police went to question him, they found him hiding in a crawl space at his retreat near the base of Mount Fuji. He was put on trial, convicted of murder, and sentenced to be hanged.

Had there been enough uranium, and time, there can be little doubt which weapon would have better suited his purposes. "Armageddon will occur at the end of this century," he once exulted to his followers. The title of one of his tracts was more specific: "A Doom Is Nearing the Land of the Rising Sun."

5

TWO RUSHES

Everything on the earth was once in the earth. Our modern world was cast and shaped out of compound arrangements of iron, silicon, copper, and carbon and more than ninety other elements. There is nothing physical here—not a single object—that cannot be reduced to one or more elements of the periodic table that once lay quiescent in the earth's crust. Mining is not the oldest profession known to man, but it is the mark of a rising civilization. Without minerals, there can be no metal, no tools, no energy, no war.

This was never truer than with uranium, the heaviest natural element, which suddenly demonstrated the power to unmake a civilization. In the middle of the twentieth century, the security of the superpowers depended on a metal that had bubbled upward long before man arrived on the scene. And now there was a race to dominate the chthonic element, wherever it lay.

There were two uranium rushes on opposite sides of the Atlantic Ocean in the 1950s. One was spurred by the government of the Soviet Union. Although the nation was officially Socialist, its drive for uranium was more dependent on free-market principles than its organizers would admit. The other rush was encouraged by the officially capitalist government of the United States. Its quest for uranium, however, relied more upon socialist concepts than its leaders wanted to discuss. And both programs left a trail of environmental and human wreckage whose effects would linger into the next century.

Both of these efforts were made in the name of atomic weapons, an ultramodern technology that happened to depend on a very old enterprise, cruder and more ancient to man even than farming: You break into the earth by force and hope to find a treasure.

The last true mineral rush in the American West began in March 1951, when the Atomic Energy Commission announced it would pay grossly inflated prices for uranium, even offering a $10,000 bonus to anybody who could develop a productive mine. It also handed out guidebooks, built supply roads, constructed ore-buying mills, and made geology reports available to anybody who had the pluck or the greed to move out to the desert and help develop a source of uranium in the American heartland in case the rail-and-sea link to Shinkolobwe was cut off.

Thousands of Americans heeded the call and rushed to the canyon country of the Southwest to look for trees that had died a hundred million years ago. The American desert had been swampy and tropical in the Jurassic era; the trees bore bizarre and riotous plumages whose green fibers were replaced almost atom for atom by the liquid uranium solutions rising up through the mudstone like a subterranean fog. The trees soaked up the uranium and lay there entombed in the innards of mesas for more than 150 million years while species above rose and reproduced and fought and fell. The inland sea evaporated and the salt domes began to wither in strange patterns, leaving behind a skeleton jumble of canyons, washes, natural amphitheaters, soaring cliffs stained dark with malachite, and sandstone pillars that resemble skyscrapers or tombstones in the moonlight, some concealing uranium-soaked logs like gold needles in their inner folds. In the shaded alcoves of some of the cliffs, a race of Indians called the Anasazi had left paintings of gazelles and misshapen humans; the people themselves had vanished in the thirteenth century.

Among the first white men to venture into the drainage system of the haunted region known as the Colorado Plateau had been the Civil War veteran and geologist John Wesley Powell, who in 1869 found "a whole land of naked rock with giant forms carved on it; cathedral-shaped buttes towering hundreds or thousands of feet, cliffs that cannot be scaled, and canyon walls that shrink the river into insignificance, with vast hollow domes and tall pinnacles and shafts set on the verge overhead and all highly colored—buff, gray, red, brown, chocolate." Winter frost had crept between the grains of the sandstone pillars and pried them apart grain by grain: a pulse of freezing and melting over tens of millions of years. The middle sections of a monolith were the weakest, and some of the centers flaked away and left a hole; these formations were called arches, and the region had more than three thousand of them among the hoodoos and hogbacks in the labyrinthine canyons radiating from the valley of the Colorado River, where a town called Moab was a cross-hatching of tidy New England rationality in a chimerical landscape.

The Mormon church president Brigham Young had ordered a few dozen families to settle the spot in 1869. He envisioned it as a beachhead for the Latter-day Saints in one of the wilder parts of southern Utah, but their hand-built fort proved vulnerable to attacks from nearby Paiute Indians, and Young quickly recalled the mission. Those first settlers didn't think much of the region, in any case. "Good for nothing, except to hold the world together," complained one. The valley was resettled in the next decade by cattle ranchers migrating over the border from Colorado who named their new postal station Moab, after a figure from the biblical book of Genesis.

This choice was curious, even for Utah, where scriptural place-names grew like sage. Here was an appellation with an odd pronunciation ("MO-ab") that happened to be derived from one of the more cryptic sex-and-blood stories of the Old Testament.

Moab was the incest-born son of Lot, who had fled with his family from the fiery destruction of Sodom and Gomorrah. After Lot's wife had been turned into a pillar of salt because she dared to look at the burning cities, Lot's eldest daughter fed him homemade wine and seduced him so the family line would not die out. The offspring of that drunken incest was Moab, who became the king of a high plain at the eastern edge of the Dead Sea that eventually took his name. The Moabites were considered wicked and fought occasional battles against Israel. Moab would also be the place where Ruth would flee to the "alien corn" in one of the most famous love stories of the Bible. But it was generally regarded as a cursed place. In Zephaniah 2:9, God declares, "Surely Moab will become like

Sodom, the Ammonites like Gomorrah—a place of weeds and salt pits, a wasteland forever."

Disgruntled locals in Utah made several petitions to the U.S. postmaster general in Washington, D.C., for a change in station name over the next decades. These requests were all denied, and the town was stuck with Moab, which was at least fitting because the ochre topography in all four directions looked similar to the Judean desert where the dramas of the Bible had played themselves out. It also happened to look beautiful on celluloid. The director John Ford happened upon the place in 1948 when looking for a picturesque desert in which to film outdoor scenes for *Wagon Master*. He loved the red-rock vistas and returned the following year to direct *Rio Grande*, starring John Wayne, who told reporters that Moab was "where God put the West." But the big game in town during the 1950s was uranium.

Moab ballooned with ex-GIs, promoters, speculators, suppliers, a few discreet prostitutes (this was pious Mormon country, after all), and assorted other fortune seekers. There was only one pay phone in town, and it was not uncommon for people to line up for the length of a block to use it. Drugstores and sporting goods stores sold Geiger counters. A store called Uranium Jewelry opened downtown. Corporations were formed over pitchers of beer at the 66 Club, one of the town's only legal bars, where, the Western memoirist Edward Abbey noted, "the smoke-dense air crackled with radioactivity and the smell of honest miners' sweat." A notorious sign was hung in the window of a brand-new drinking establishment, the Uranium Club: NO TALK UNDER $1,000,000. The schoolhouse was jammed: Classes had to meet on the lawn outside, and the school felt obliged to start offering free lunches because so many parents were off prospecting in the desert.

A man named Joe Blosser told a reporter he had abandoned a pleasant life of golf and cocktails in California to join the hunt in the desert. "I guess it's freedom I want, and the sense of being useful," he said. "This stuff uranium . . . We need it. It's a new kind of power. There was coal to make steam, oil to run gas engines. Now there's uranium."

Another of the migrants was an owl-eyed Texan named Charlie Steen.

He had grown up in the backwater town of Caddo, the son of a riches-to-rags oil prospector. Steen worked his way through college, learned a little Spanish, and landed a job with Socony Vacuum Oil Company after graduation. Sent to Peru to look for new prospects, he came up empty. Far from discouraging him, the experience left him determined to start hunting wealth on his own. But he had a wife to take care of, so he took an exploration job with Standard Oil of Indiana. After an argument with two of his supervisors, he was fired for being "innately rebellious against authority." Steen was working as a carpenter in Houston in 1949, plotting his next move, when he happened to read an article in the *Engineering and Mining Journal* about the desert that straddled the Colorado-Utah border. CAN URANIUM MINING PAY? asked the headline. Steen decided he would try.

He drove into Moab with a jeep and a trailer, his wife and four children; a wispy-framed striver with dun trousers and a grin shaped like a wedge of orange. His receding hair and thick-framed glasses gave him the air of a NASA technician. And indeed, he was one of the few prospectors arriving on the Colorado Plateau who had graduate-level training in geology. Steen had a theory about these formations that others thought was foolish. He believed that the same method used to find oil deposits could be used to find uranium. He spent almost no time looking for the easy money in the canyon walls and he didn't even own a Geiger counter.

The key for Steen was the anticline, a dome-shaped structure some distance *behind* a spot where a trace amount of uranium had already been found. He believed that the first trace could be the petrified whiff of a larger deposit hiding beneath the clay. This was one of the guiding principles of oil exploration, a field he understood and where he felt most comfortable. A bore of two hundred feet would find a patch in the anticline, if it was there. Others drilled horizontally into cliffs; Steen would bore straight down into the soil. This contradicted everything that was known at the time about uranium geology, but Steen was cocksure.

"Turn a spoon upside down on the table, like so," he later explained. "That's your anticline. Down the flank of that dome, down below the rimrock, that's where the uranium is. That's where the oil often is."

Armed with a $300 grubstake, he began to focus his attention on the Lisbon Valley, about thirty miles south of Moab. The Atomic Energy Commission had found low-grade uranium in an outcropping on the west side of the valley, but the uranium was judged too high in lime content to make any exploration profitable. This seemed like the perfect place to test the anticline theory. Steen staked out a dozen claims on the valley floor and gave them wistful Spanish names to remind him of better days in Peru; Mi Corazón ("My Heart"), Linda Mujer ("Pretty Woman"), Te Quiero ("I Love You"), and Mi Vida ("My Life"). This last choice of name, according to Edward Abbey, demonstrated "revealing pathos." At the time, Steen had that patch of desert nearly to himself. The AEC didn't think much of these barren spots, located in a district called Big Indian, and one official called him a "crackpot."

By this time, he and his wife and their four children had spent all their savings and had to sell their trailer. They were living on corn bread, venison, and beans inside a $15-a-month tar-paper shack in the forlorn railroad village of Cisco. Time and money were running out. In the field, Steen ate only potato chips and mushy bananas. He could no longer afford new boots, and his toes stuck out of the ends of the ones he had. His wife could satisfy her nicotine cravings only by picking up half-smoked cigarettes on the side of the highway and rerolling the tobacco.

The birthplace of uranium mining, St. Joachimsthal, meanwhile, was in for special misery.

The Red Army had been sent to capture the medieval mountain town shortly after Hitler's suicide, and geologists went into the shafts to assess the quality of the pitchblende. What they found pleased them. "It was not the Soviet Union that decided that the atomic bomb was going to be the weapon of the future," reasoned one colonel.

In November 1945, Joseph Stalin pressured the government in Prague into signing a confidential treaty: Moscow would get the entire run-of-mine, and Czechoslovakia would provide the labor force under a state corporation named grandly—and vaguely—National Enterprise Jachymov.*

---

* Jachymov is the Czech name for the town.

The mines that had worried Albert Einstein were now secretly under the control of the Russians.

Czech foreign minister Jan Masaryk was left uninformed of this deal when he promised the United Nations that St. Joachimsthal's uranium would never be used for "mass destruction." He had added: "We in Czechoslovakia want our uranium to be used entirely differently—to build, protect, and make our lives safer and more efficient." He had concluded with a proposal that echoed the failed Baruch Plan: that an international organization should inspect all uranium mines to ensure that none of the ore would be used for weapons.

This speech helped sign Masaryk's death warrant. Nobody had briefed him on the exact details of the secret uranium deal with the Russians, but he was reprimanded. Two weeks after the Communists seized power in Prague in 1948, he was found dead, lying in his pajamas below a bathroom window in the courtyard of the Foreign Ministry. The official verdict was suicide, but Masaryk was widely believed to have been murdered.

Stalin ordered the Czechoslovakian mines rushed back into production, but there was almost nobody around to do the work. The town was nearly deserted. The German-speaking residents of St. Joachimsthal who hadn't been drafted into the army had been expelled from the region after the Nazi defeat. The entire motor pool consisted of a truck and two horse-drawn wagons. The Russians assigned some of their prisoners of war to start digging uranium, and when that proved too slow, they turned toward the ordinary citizens of Czechoslovakia.

The minister of justice, an ardent Communist, told his subordinates that "we must concentrate all our attention" on the labor problem and round up able-bodied men wherever they could be found. A directive called Plan of Action T-43 was issued; it contained this sentence: "We need 3,000 more people who do not already work for us." Neighbors were encouraged to inform upon one another for petty crimes. The state security bureau, the SNB, was also given latitude to detain suspects on vague ideological grounds and bring them in front of special committees that, if there was no genuine crime, declared them guilty of such things as "believing in bourgeois ideas" or being a "product of a capitalist milieu." These prisoners were described with an antique term: *nevolna,* or "serfs." They were the first residents of what would become a giant uranium gulag.

The day of their arrival at St. Joachimsthal was a shocking and miserable experience. The hills had become dotted with more than a dozen new crossbeamed headframes, each one encircled by two layers of barbed-wire fences and watched over by four elevated guard towers, manned by Russian soldiers holding carbines. The space in between the fences was a moat of fine white sand, raked regularly so that any footprints would be immediately apparent. A large red star was mounted over the main gate to each camp. Loudspeakers mounted on the fences blared patriotic music and speeches.

The prisoners were marched inside the gates, handcuffed to one another in a line, and made to stand at attention for their first roll call, a thrice-daily ritual they would be made to repeat, in every kind of weather, for years to come. Anyone who stepped out of line was beaten over the head, usually with a giant ring of keys, and their bloodied foreheads served as a warning to others. "It was a nightmare. I cannot believe that this system was designed by Czech people," one inmate recalled in a letter home. He added: "It looks like *arbeit macht frei*," referring to the infamous sign above the Auschwitz camp that meant "Work makes you free."

Among those rounded up was Frantisek Sedivy, an idealistic twenty-five-year-old vocational school student who took part in an underground movement opposed to the Soviet-backed regime. He and his friends had already risked imprisonment by helping two families smuggle themselves to West Germany. Sedivy's group was approached by a man who wanted to arrange more defections. Sedivy was cautious, but agreed to help. He was promptly arrested and told that his new "friend" had actually been a police informant conducting a sting operation. Sedivy was sentenced to fourteen years in the uranium mines.

"It was not forced labor," he would say later. "It was slave labor."

The daily ration of food was four slices of bread and a few mushy vegetables. Three times a week came watery "soup," which was lukewarm water added to dried vegetables. Within a few months, Sedivy lost nearly forty pounds. Snowstorms came to the mountains in October and lasted intermittently until April; the drifts blew through the open windows of the barracks and onto the cots. Beatings were common. When jingoistic speeches were not being trumpeted on the loudspeakers, the administrators played the same three Czech folk music records over and over.

A favored job was minding the kennels, where a prisoner might hope to steal a bit of dog food every now and then. But most prisoners were ordered into the mines. Sedivy's first assignment was inside the Svornost ("Concord") shaft, near the center of St. Joachimsthal, which had first been excavated by silver-hungry peasants in the sixteenth century. There was no training: Men were simply handed pneumatic drills and ordered to bore holes and lay track. The tunnels were barely wide enough to accommodate the ore carts; men had to flatten themselves against the walls when one came hurtling by. Carts also jumped the tracks. "We had a lot of broken legs," recalled Sedivy. He was given a rubber coat, which kept him partially dry from the moisture oozing from the walls. The water could rise to the knees before pumps kicked in. On bad days, workers came out of the elevator cages so drenched in mud that friends could not recognize one another. Almost nobody was issued a helmet, and workers were routinely killed when the explosive charges sent chunks of rock whizzing through the tunnels.

Sedivy later worked in the crushing mill, a primitive facility where inmates broke up rock with sledgehammers. His job was to help separate chunks of pitchblende—called *smolinec* in Czech—according to purity. The lowest-quality ore was set aside for the waste piles. Medium- to good-quality ore was packed into crates, which weighed approximately 150 pounds each. These were loaded onto railcars for shipping to the Soviet Union. The mill produced nearly 270 tons on a good day, and its windows were frosted gray with dust. The four-story building became known as the "tower of death" by some miners—including Sedivy—who had heard there was something dangerous about breathing uranium particulate but were powerless to do anything about it. There was no doubt in the mill about what the pitchblende was going to be used for, although it was rarely discussed.

"We knew that the Soviets wanted the uranium for bombs," said Sedivy. "There could have been no other purpose. It wasn't for science."

He dreamed of escape, but the penalty for being caught outside the fences was death on the spot. Bedsheets were issued only in summer months to prevent the inmates from using them as cloaks to blend in with the snow. Sedivy recalled being forced to march in a circle around a pile of corpses. They were men who had tried to sneak away and had all been shot in the face, their identities obliterated. Their families were notified with a curt letter. A typical two-sentence notice, addressed to a woman named Anna Cervenkova, said: "Your brother, Petru Frantisek, was shot in an escape attempt. You are not permitted to attend the funeral." There was no signature.

Sedivy recalled a rare moment of levity shortly after Stalin died of a stroke in March 1953. A political officer in the camps was delivering a windy eulogy inside the barracks and, at one point, tried to refer to Stalin by his full Georgian birth name, Iosif Vissarionovich Dzhugashvili. The officer's Czech pronunciation was poor: To the inmates, the mangled name came out sounding like "Joseph has fleas and syphilis." There was nervous chuckling, and that was the last time that that officer was allowed to make a speech. At about the same time, rumors had spread of the election of General Dwight Eisenhower to the office of U.S. president. This inspired speculation that the same general who had fought the Nazis would also initiate a war against the USSR. This was cause for some hope.

A miner who managed to escape told the following story to a Western journalist. "Once in a while something happened that encouraged you to carry on," he said. "The SNB guards had a beautiful, expertly trained police dog that was said to be worth forty thousand koruny. The miners were fascinated with the animal. Everybody agreed that he would make a fine roast. One night, three miners came upon the dog when no SNB man was around. Poor dog; his expert training didn't help him. They lured him into an abandoned mine shaft and killed, skinned, and cooked him on the spot."

Inmates did their best to cheer one another up amid the boredom and despair, sharing tobacco and new coats when they were available. "The people here are tough and brave," wrote Viktor Opavsky in a letter home to his family. He mentioned in particular one Father Harman, a Catholic priest, who tried to keep his fellow inmates from falling into depression. The letter, which would have been read by censors, does not indicate if Father Harman continued to say Mass in secret. In any case, priests were the targets of special harassment in the camps. The only rations of meat in the week were often served just on Fridays to make them choose between their consciences or their stomachs. The Bible and other books were forbidden, but some prisoners managed to have them smuggled into the camp, where they were hidden underneath rocks and behind barrack walls. Discovery meant punishment, usually confinement in a freezing underground bunker for a month or more. Serious violators were beaten with rubber hoses or hung from metal grills for hours. One guard earned himself the nickname the "Human Beast" for ordering prisoners to stand outside during winter storms and shoveling more snow around their feet as it blew away.

Those who worked the hardest were given the best equipment and allowed to skip the dangerous tasks. They were permitted to watch movies on Sunday and could go into town by themselves for brief periods, almost as if free men. Early parole was also offered. These elite brigades were known as Stakhanovites, after a workaholic miner named Aleksei Stakhanov, who had mined more than one hundred tons of coal in six hours in 1935 and become a symbol for Socialist excellence. Those who joined often embraced the propaganda even more passionately than their teachers.

"If you entered one of their barracks, you would be under the impression that you had entered a Communist Party indoctrination classroom," said one inmate. "You would see red banners all over the place and political slogans taken from the Communist daily press, which would convince you that the people living there did not participate in the Stakhanovite movement just because it is the only way out."

The gulag labor force was supplemented with regular wage-earning mine workers, who were free to visit the main street of St. Joachimsthal after their shifts. "On Saturday nights, the place rolls and rocks like a

Klondike shantytown, blessed with a particularly rich strike," said one contemporaneous account. "Cheap music, cheap liquor, and cheap women abound." The saying was that there were three shifts at St. Joachimsthal—working, sleeping, and drinking.

The paid miners were the only ones who could enjoy it. The valley had become a security zone—nobody was allowed in or out without papers. Military checkpoints and barbed wire were set up at the bottlenecked entrance. The spa resorts became barracks for the Russians; one notorious unit interrogated suspects in the cells of a former nunnery. By the autumn of 1953, more than 16,100 inmates were being forced to dig, crush, and load uranium at St. Joachimsthal. More than half this number had been jailed on political charges. But this would represent the population apex for the uranium camps in Czechoslovakia. The mad thirst for uranium had dissipated somewhat after Stalin's death. By then, the Soviet nuclear program had developed an estimated 120 atomic bombs. Uranium had also been located and mined in the Urals and in the secure regions of Kazakhstan, and production on the German side of the mountains was also outpacing Czechoslovakian production by nearly six times. Prisoners who completed their sentences were generally not replaced with new ones.

The town had earned the nickname Jachymov Hell, and the valley that had given its name to the American dollar had been choked with tailings piles. Pyramids of waste, gently fuming with radon gas, were miniatures of the medieval hills that brooded over them in all directions. "The western part of the town disappeared under the waste banks," noted a local historian, "and the valley along the banks of the Klinovec brook with the playground, swimming pool, and vacation restaurant were buried under the waste."

Frantisek Sedivy was transferred to a new set of uranium diggings near the town of Pribram. Conditions there were markedly better. Most of the laborers were regular wage hands, and the free men shared food, warm clothes, and cigarettes with their incarcerated companions. They could also be persuaded to smuggle uncensored letters in and out. Inmates also worked out a kickback scheme with their free colleagues—if a prisoner happened to discover pitchblende, he would pretend as though the free man had found it; the resulting bonus would then be quietly split between them.

Sedivy was released from the uranium camps in 1964, having served a total of twelve years. There was nobody to welcome him home when he returned. His mother and father had died when he was in the camps, and his girlfriend at the vocational school had married another man a long time ago. Sedivy set about the task of restarting life. He found work as a welder and went back to school to earn an economics degree. Later, he worked for a book publisher and then for an agency that ensured the safety of roads and bridges.

I met with him in the back room of a tavern in the rural town of Revnice, where he had retired after his final government job. He was eighty years old at the time, but had a face free of wrinkles and a checkered tweed jacket pressed clean. His hair was still mostly black, though striped with white in the middle.

He told me about his years at St. Joachimsthal in a calm and untroubled voice, a nonalcoholic beer barely touched on the table before him. Almost all his friends from the camps are dead, mainly from lung cancer, but some from other ailments. Tobacco had been a major cash commodity in the barracks; it was one of the only little pleasures in an otherwise dreary existence, and it was usually smoked inside a roll of newsprint, for lack of any other paper. Sedivy never developed a taste for these improvised cigarettes. It may have saved his life, for he never showed any signs of cancer.

He lives in retirement with his wife and adult son and has tried to let go of bitterness.

"I don't hate uranium," he told me. "I don't even hate the Russians. They were mostly simple people, only doing what they were told. They would go to the gulag themselves if they hadn't made us work. . . . They wanted their *smolinec*. That was all."

In the narrative of American mineral exploration, few myths are as cherished as that of the busted prospector, discouraged and about to quit, who sinks a final hole only to happen on the strike of a lifetime.

A man named Alvinza Hayward, for example, was said to have been obsessed with a gold claim he had purchased in Amador County, California, in 1856. Indicator minerals were present in the samples, but the ore itself was too poor to mill and got no better as the shaft deepened. Most of his crew eventually walked off the job, believing the enterprise to be hopeless. Hayward's friends started to desert him after he begged them for more money. Only one hopeful rancher could be persuaded to front a final grubstake. Hayward burned through his last friend's $3,000 and still there was no gold. Near the end, bankrupt and despairing, he claimed to have been unable to buy a new pick and had eaten his way down to a bag of dried beans. That was when he intersected the main ore body. The Hayward Mine became the most lucrative in that part of the Sierras and earned $5 million for its indefatigable owner.

A parallel legend is associated with the Enterprise Mine of southwestern Colorado. A luckless prospector named Dick Swickheimer had borrowed heavily to dig a shaft on Dolores Mountain and seemed headed for ruin when a winning lottery ticket gave him the money he needed to sink his borehole a few feet deeper. And then: silver.

In Arizona in 1877, U.S. Army officers had warned the ragged wanderer Ed Shieffelin that "all you find out there will be your tombstone" when he told them of his intention to explore some hills inhabited by hostile Apaches. He instead found the top of a silver vein protruding from the ledge of an arroyo; the unruly town that sprang up around it was named Tombstone.

No less a figure of frontier mining than George Hearst, the father and bankroller of William Randolph Hearst, told a story about nearly quitting on the side of a trail in 1859, just before he made his fortune at the famous Comstock Lode in Nevada. "I felt old and used up and no good," he told the *San Francisco Mining and Scientific Press*. "My sense told me to turn back and make my fight where I was known. There was safety in that anyhow. But I'd been camping night after night with the boys ahead of me, and it made me lonesome to think of parting company with them. So after switching and switching the dust on the trail and feeling weak and human because I yielded, I mounted my horse and went after the party. I got to the Comstock, and in six months

I'd made half a million dollars. That was the foundation of what I've done since."

These stories are part of a body of mining lore called discovery tales, which tend to be attached to substantial finds. Other common themes include the prospector being shown where to dig by a tribe of local Indians, seeing the landscape in a dream, or accidentally kicking over a stone in the middle of nowhere only to find it flecked with precious trace metal. A big strike seems to demand a romantic backstory. Perhaps the reality of most mineral exploration—plenty of rote physical labor and monotonous data analysis—seems unequal to the grandeur of the unearthed treasure, and the discovery tale becomes a means of crediting more ethereal forces of destiny or Providence or ancient aboriginal wisdom.

What the stories gloss over, however, is the necessary element of manic depression associated with mineral exploration, which was—and is—an endeavor where the vast majority of propositions end up in failure, where financing is always precarious, and where the few schemes that succeed do so because of the generally irrational faith of the central actor.

Geology is an inherently mysterious business; what lies underground is a matter of deduction and inference. Enormous amounts of energy and capital must be expended in the cause of theory. Such a profession tends to attract the gambling personality, a man willing to stake his reputation and livelihood on a hunch. Risk is the drug. Failure brings only temporary depression, soon to be replaced with the refreshed insanity of staking new claims. There are always more holes to drill, more investors to tap. Millionaires who claim to have been on the verge of quitting when they found a jackpot had simply been living normally; their material salvation most likely arrived on an otherwise unremarkable day. The brink of failure had always been a comfortable place for them to pitch a tent.

In any event, the American uranium rush was to have its own piece of nick-of-time mythology in the person of Charlie Steen, who had been obsessed with the idea that uranium could be found in an anticline formation—a theory that others thought was nonsense.

In July 1952, broke and discouraged, Steen took a diamond drill out to his Mi Vida site, where the walls of a canyon parted like a pair of outstretched legs, and managed to dig out some multicolored core samples down to 197 feet. Then, on July 27, the drill bit broke off the pipe and got lodged in the hole. His core samples seemed worthless. There was the usual deep red clay and some grayish rock that looked like frozen tar, but none of the yellow carnotite he was seeking.

Steen tossed the pieces into the back of his jeep and drove back to Cisco in a bleak frame of mind. He was nearly out of cash. Before going to tell his family the bad news, he stopped at a service station near his tar-paper shack. Steen's son Mark would later write that his father was on his way to Grand Junction, Colorado, for new equipment and that he had every intention of continuing to drill. But his mood on that summer evening was one of despair. The owner of the station, Buddy Cowger, had a "Lucky Strike" Geiger counter and, as a grim joke, Steen asked him to wave it over the useless cores in the back of his jeep. The reddish sandstone showed nothing, but the gray cores sent the meter's needle all the way to the edge. The dingy rock turned out to be uranitite, otherwise known as pitchblende. Steen had never seen this particular oxide outside of a museum. But he had just tapped into a huge vein of it. Steen whooped all the way to his shack, nearly decapitating himself on a clothesline. "It's a million dollar lick!" he yelled to his wife.

In 1953, the same year that the slave labor force reached its apex in Czechoslovakia, Charlie Steen became fantastically rich overnight and built himself a space-age mansion atop a mesa on the north edge of Moab. He named it Mi Vida—"My Life," after his mine—and then proceeded to memorialize almost all aspects of his former poverty. He bronzed the worn-out boots he was wearing on the day he found the uranium and displayed them on a mantel in his house. He had his lantern plated in gold. In a public act of sweet revenge, he bought up shares in a bank in Dove Creek, Colorado, that had refused to lend him $250 when he was poor. "Don't ever bounce a prospector," he told reporters. "He might come back someday and buy the bank." When he won $10,000 with four of a kind in a poker game, he had an oil painting commissioned to commemorate the lucky hand. He was written up in *Newsweek*, *Time*, *BusinessWeek*, and *Woman's Home Companion*, bringing more luster to the uranium effort than the AEC ever bargained for.

"I couldn't have been more delighted because he was one of our first millionaires," an agency official told the journalist Raye Ringholz. "That was what we needed . . . that flair, publicity attached to someone who was on his uppers. We need guys like that. He's a departure from the norm and that's the kind of guys that civilization makes advances on. We're not going to make much progress with the ordinary pedestrian-type individual."

Steen used to complain he had to spend more time hunting grubstakes than uranium. Now he seemed to be occupied primarily with hunting headlines. "One of the things that kept him busy was seeking publicity," recalled his onetime partner Dan O'Laurie, who was squeezed out of Steen's Utex Exploration Company early on. "He liked publicity, good or bad. It didn't make any difference, just as long as it was publicity. That was the nature of the man." Steen booked speaking engagements around the country, telling audiences about the future of nuclear power and his own role in helping build it up. "Moab will never go back to what it was," he told one audience. "Atomic bombs have saved more lives than they destroyed." The Pentagon rewarded him with a top-level "Q" security clearance and warned him not to disclose the size of his ore reserves, lest it give the Soviets a picture of American strength.

At his hilltop mansion, he threw parties such as Utah had never seen before, with free-flowing champagne and oysters flown in from Maine. Henry Fonda and Anthony Quinn were among those invited for dinner. The home resembled a version of a tract house from Levittown, only one that had been inflated to twice the normal size and teleported to the top of a rocky sandstone precipice. The clocks on the fireplace mantel were fixed at 5:05—to celebrate the permanent cocktail hour, Steen explained—and his poolside deck featured a commanding view of the unfolding boomtown of Moab below where the gaslights of a hundred roughneck housetrailers glowed at night. Those who weren't invited to the parties sometimes hiked the serpentine driveway to crash in. A running joke, according to local historian Maxine Newell, was to dunk guests in the pool. Steen himself took weekly chartered flights to Salt Lake City to a dance studio where he was learning the rumba. When there was a television program he wanted to watch one evening—Jackie

Cooper portraying him in a live drama called "I Found Sixty Million Dollars"—he went out to his private plane, which was equipped with a television, and ordered his pilot to fly around in circles above the mesas where the signal was stronger.

Another uranium celebrity was Paddy Martinez, a Navajo shepherd who lived in a hogan outside the highway town of Grants, New Mexico. He had been tending sheep atop a peak in full view of Route 66 below when he got a hankering for a smoke. "I was on horseback, going along a trail to Rattlesnake Trading Post for the cigarettes, when I saw this little yellow spot under some rock," he told a journalist. "I dug it out with a stick because it reminded me of the time in 1947 when I bought a bus ticket in Grants at the Yucca Hotel. Three white men were talking about an ore called uranium and saying it was worth a lot of money. They were showing some of it to each other and I got a look at it. It was the same yellow stuff I was holding in my hand on that trail. Well, I got my cigarettes and came home and told my wife I'd found some kind of ore. She didn't believe me."

The rock was carnotite, and Martinez had found a patch of it big enough that five different mills were necessary to process it all. The population of Grants tripled in three years. The claim Martinez staked turned out to be on the property of the Santa Fe Railroad, however, and he received only a finder's fee, paid to him in monthly stipends of $250.

"This damn uranium," he complained. "My friends don't like me the same anymore."

A man who saw a better payday was Vernon Pick, an affable Minnesota electrician who claimed to have wandered into a field of carnotite in a desolate patch of desert near the base of a towering ledge called San Rafael Reef. Almost out of food and delirious from drinking arsenic-laced water, he built a raft of driftwood and floated his way down Utah's Muddy River to report his claim. Critics later maintained that a pair of colluding Atomic Energy Commission agents had tipped him off as to where the ore could be found, hoping for a kickback. The FBI investigated. No criminal charges were brought. But the mythology was too good to resist. In *Life* magazine the nation read the gripping story about a simple man who overcame the wilderness. "He fought storms, rattlers, poison, death itself to find a uranium bonanza!" enthused the subtitle. As a discovery story, it was first rate, even if it probably wasn't true. Pick sold his mine, which he had named the Hidden Splendor, for $9 million and a used airplane. It seemed that the hope of America was once again on the Western frontier and buried underground, where anybody could come and find it.

A prospector's basic gear included a sharp-pointed pick, a pair of binoculars, a jeep or a mule, a "Lucky Strike" Geiger counter, and a battery-operated ultraviolet lamp that made radioactive rocks glow. The seekers would spend weeks alone in the desert, hiking the blistering wastes and eyeing the cliffs for the hints that might give away a prehistoric stream with some fossilized trees buried within. These could usually be found in the Shinarump layer of sandstone, which was like a wedge of crunchy pink mortar between the Chinle and the Moenkopi formations.

"It's a little different from other mining because it's dried vegetation, animal life, trees," said a driller named Oren Zufelt. "One time we followed a vein of ore down what looked like a little creek and some timber had fallen in it. We worked down into the rocks and followed that vein up around the rocks and there was this dinosaur. He had jumped into this mud and got stuck, died, and made uranium. It was just like he had fallen in there yesterday."

Some of the uranium-soaked trees, known as trash pockets, lay so shallow that all a finder had to do was shovel the ore into a burlap bag and load it onto a mule. The ore itself was yellow or gray, but it was known to have wildly prismatic effects on the rocks it adjoined.

"I saw a man once find a whole tree that was just high-grade uranium. It came off like black powder, really soft, like pepper," recalled Jerry Anderson, who prospected with his father. "It was a tree about two feet in thickness and the branches went off maybe ten feet in each direction. He blasted that thing out of there and got sixty-five hundred dollars. When the tree was depleted, that was the end of his uranium ore."

Even little old ladies could get into the game. Edna Ekker was the matriarch of a ranching family whose roots traced to the Utah frontier; as a young girl, she had once accepted a piggyback ride from the bank robber Butch Cassidy, who had stopped by to hide a bag of money in the fruit cellar. She quit raising horses in her seventies and turned to uranium, which practically could be picked up off the ground with a few turns of the shovel. "It was really shallow work where we were," she told an interviewer. These surface diggings, small enough to be worked by one or two people, were known as dog holes, and they began to sprout up off the dirt roads. Prospectors marked these claims by building cairns and writing their names on forms they sealed inside tin cans—usually a Prince Edward tobacco can—at the base. They liked to bestow fanciful names on their mines: Bull's-eye, Black Hat, Whirlwind, Hideout, Royal Flush, S.O.B., Payday.

The richer prospectors searched with helicopters or airplanes. Jerry Anderson's father rigged up an ultrasensitive scintillation counter that hung from the tail of his light aircraft as he cruised through canyons and past mesas. His sons were told to drop bags of powdered lime or flour on top of anything promising. They would hike in later to find the white mess and examine the spot up close. Paranoia ran so high that some of the company pilots refused to make a second pass over an area that registered a spike, according to the historian David Lavender. They were afraid somebody would be watching them through binoculars and barrel in to claim that ground before the plane could even land. A story made the rounds about a haul truck that dumped its load of uranium ore on the highway. Somebody quickly filed a claim on the highway.

Those who wanted to do serious mining had to hire a bulldozer to come plow a road to their site. Any kind of dirt track would do, and the roads were an awful kind of art in themselves. They were blasted into cliff sides, routed through arroyos and up makeshift dug ways, and carved into precipitous slopes at terrifying gradients. "Some of the ecologists today would be very unhappy with me if they saw some of the roads we built," laughed a former Atomic Energy Commission official in 1970. Several truckers lost their lives in spillovers. One unfortunate man on the remote White Rim Road was on his bulldozer when it tipped over and pinned his arm to the ground just after a supply truck left. Nobody would be back for another three days; the heat and thirst surely would have killed him had he not fished out his pocketknife and sawed off his own arm.

After a prospector had his road built, it was only a matter of getting at the fossilized trees. The strategy did not differ from the blast-and-tunnel methods perfected in Appalachian coal mines in the previous century. Hammers and dynamite were used to bore into the mesas. The debris was shoveled out by low-paid muckers—often local Navajo Indians—who loaded the rock into mine-track cars or wheelbarrows. The sandstone was usually sturdy enough to tolerate the passage of a mine tunnel, but the passages had to be braced with timbers. Miners bragged about their ability to "smell" the uranium inside the earth.

Charlie Steen did his part to fuel the boom by opening a hangar-size mill that at its peak employed nearly three hundred workers and sixty truckers who hauled in the ore from all points on the Colorado Plateau. Here the ore was pounded into gravel by a series of steel plates, roasted, and then mixed with sulfuric acid or sodium chlorate to leach out the uranium-containing portion. The resulting solution was run through an ion exchanger—a device like a conventional water softener—to remove the sickly residue of uranium oxide, known as yellowcake in the industry and as "baby shit" in the vernacular of mill employees. Most of them would moonlight to hunt for their own versions of Mi Vida in the ghostly red-rock desert, which featured more than five hundred working mines of varying sizes. And the immigrants kept coming.

"There were lots of camp trailers down by the river, and every sign in town had the words 'atomic' or 'uranium' shouting from the stencils," recalled Tom McCourt. "But if you looked behind the facade, Moab was still wearing the overalls and straw hat of her agricultural Mormon founders. There were always farmers and cowboys in town, and beat-up old pickup trucks with hay bales and ugly dogs in the back."

Charlie Steen bought the old Starbuck Motel building and turned it into workers' quarters. And when that grew full, he started construction on a new subdivision of detached family homes named Steenville. He could afford it. Within two years, the Mi Vida mine and his Utex Exploration Company were worth $150 million. Never a particularly good businessman, Steen got into a tangle of lawsuits with former partners and people he accused of "claim jumping" plots of land near Mi Vida. At one point, he grew frustrated with his board of directors and abruptly decided to fire them all. "The way he did this," recalled his metallurgical superintendent Clement K. Chase, "was to get someone from Grand Junction to come down and change all the locks at the office so nobody could get in the next day." But he was also ridiculously generous, donating money for schools, churches, and college scholarships, and to just about anybody who asked him. Recalled his wife, Minnie Lee, to an interviewer, "All they had to do was pat him on the back and tell him how great he was and say, 'Can you let me have about fifty thousand dollars, Charlie?' He loved them all." Before long, Steen acquired the inevitable nickname "Good-Time Charlie." On a family vacation to Spain, he rented out an entire carnival for the night so the poor children of the village where he had been staying could enjoy Ferris wheel rides and cotton candy.

"Maybe I'm crazy," Steen told a reporter. "Maybe I should sell out and get out. But I don't want to retire with a bundle. I like this life." He was then just thirty-three years old.

The northern slope of the Cruel Mountains was in Germany, where the range was called by a different name: the Ore Mountains.

Mining had been the backbone of the economy here for eight hundred years. A local legend said that the first wagons to pass through here showed traces of lead on the wheels, which started a Dark Ages mining rush. The folds and valleys became dotted with plump little parish towns of cobblestone and steep roofs. Pits that had been first dug in the twelfth century yielded silver and tin and were named with the Bible and church history in mind: King David, White Dove, St. John. The mountains were also famous for the manufacture of lace cloth and the sale of nutcrackers that miners carved during the boring winter months. They were shaped like dolls and dressed to resemble German politicians.

The miners of the Ore Mountains considered themselves a tough and capable breed, even when they were unemployed, and hailed one another with the greeting "*Glück auf*," a phrase that literally means "Luck up," but is more akin to "Luck is coming to you," or "Good luck on your way up." Their swagger was irritating to some of their neighbors in Saxony, who called them "shaft shitters" for their supposed penchant for defecating underground.

Their silver was also mingled with uranium, considered an irritant up until the 1920s, when leading citizens of the town of Schlema sought to cash in on the craze for radium water and spa vacations. They sank twelve wells and built the grandiose Radiumbad Oberschlema Hotel to lure the moneyed set down from Berlin, a convenient four hours away by express train.

There were bathing pools and cafés and gambling halls; all the things that a Weimar-era holiday demanded. A photograph from 1939 shows dapper guests lying in reclining chairs, breathing through cone-shaped masks that hissed with radium steam. It was thought this was healthy for the lungs. Jews were pointedly barred after Hitler came to power, but the spa continued to do good business through World War II, so good even that approximately two thousand customers came to inhale radioactive steam even in the disastrous year of 1945, when most of Germany's cities lay in bomb-cratered ruin.

The Red Army moved into Schlema shortly after the war, as it had into St. Joachimsthal, and the geologists were stunned at what they found: veins of pitchblende up to eleven yards wide. These were some of the richest deposits ever found outside of Africa. The Russians quickly formed a state mining company with a deceptive name: Wismut, which means "bismuth." And for a director, Stalin picked a man who was already hated and feared.

General Mikhail Maltsev, the son of an electrician, had overseen a network of forced-labor camps and coal mines in the freezing Vorkuta district of Siberia during the 1930s. His ruthlessness against dissent was already legendary, as was his loyalty to the regime. Maltsev was now transferred to Germany and put in charge of supplying uranium for the world's second all-out program to build an atomic bomb.

In his own way, Maltsev was as driven and as effective as Leslie Groves had been in kick-starting the Manhattan Project. His main obstacle was a lack of manpower, so he employed hiring tactics that were familiar to him from Siberia: slave labor. It was already being used across the mountains in St. Joachimsthal.

"We are Bolsheviks," Maltsev told a meeting of party officials, "and there is no fortress we cannot storm." Police were ordered to round up drunks in bars and vagrants in railroad stations. When that source ran dry, ordinary people—taxi drivers, schoolteachers, butchers, pharmacists, and waiters—were convicted of phony crimes and deported to the radium towns. One of the important early recruits was a mining director named Schmidt, an enthusiastic Nazi who had been detained in a prison camp after the war. Once the Soviet authorities realized he had impeccable knowledge of the tunnels, he was recalled to the uranium mines and told he would "pay with his life" if production was not boosted immediately. Schmidt hardly slept from then on, and spent all his waking hours on the job, cheerfully showing up in the middle of the night if he was called in for even a minor problem. The Russians rewarded him with a new apartment and a car, and he would go on to be as passionate a defender of Stalin as he had been of Hitler.

Maltsev organized his unskilled draftees into brigades of six men each. They were handed shovels and pneumatic drills and ordered into the tunnels first dug by merchant burghers in a distant century. The tunnels were quickly widened, extended, and rail tracked by swarms of emaciated German refugees. The genteel Radiumbad Oberschlema Hotel was razed after uranium was discovered underneath, a fate that eventually befell almost the entirety of downtown. Heaps of waste rocks were pyramided where the houses used to be; the surrounding hills came to resemble a volcanic desert.

Those too sick to work underground were handed Geiger counters and told to comb the birch and pine forests for any sign of radioactivity. The hills quickly became scarred with dog-hole excavations. Otherwise valuable minerals such as cobalt and bismuth were treated as nuisances and abandoned; exploiting them for their own sake was considered too time-consuming. Uranium had become the only pursuit. "Ask for what you like," Stalin had told his nuclear scientists. "You won't be refused." What they needed most of all was uranium, and meeting the preassigned quotas was now a matter of life or death for those who had been sentenced to work at Wismut. The company slogan, "Uranium Every Hour," was now alpha and omega, mantra and meaning.

Maltsev drove his prisoners hard. The powerful *rat-tat-tat* of the pneumatic drills—known in miners' patois as shooting tools—wrecked the nerves and sinews of men not accustomed to hard industrial labor.

The heat and claustrophobia of the mine shafts were overpowering for some. In the shafts heated by natural radium water, the temperatures were known to reach 120 degrees Fahrenheit, even in winter. Most men assigned to work in the deep adits stripped down to their undershorts to keep from passing out. A former barber named Heinz Pickert, who had served on a U-boat during the war, worked himself so hard that his hands grew numb and began to shake at all hours of the day. He tried to go back to work as a barber for Wismut, but he had to quit because his jittering hands could no longer hold a pair of scissors.

The rail lines leading out of the mountains had been damaged by Allied bombs, and so the first good chunks of pitchblende had to be loaded into the trunks of cars and driven to an air base in Dresden, where they were flown into Russia. On Joseph Stalin's sixty-ninth birthday, some officers tied a large red bow around a large piece of pitchblende as a "present" for the general secretary.

American intelligence agencies had only a hazy idea of what was happening. A network of enrichment plants had been built in Siberia, but their locations were hard to pinpoint. The emergence of new uranium mines in Russia's Ural Mountains was overlooked by Western intelligence agencies, still operating under Leslie Groves's assumption that uranium was a rare resource. A U.S. intelligence report from 1946 estimated that the Soviet Union itself would produce no more than two hundred tons per year over the course of the next four years, a badly flawed estimate. The U.S. consulate in West Berlin offered cash rewards for any defector from the uranium mines who could bring in a few lumps of the ore so the purity of the enemy's reserves could be analyzed.

When news of this reached Maltsev, tougher security procedures were put in place. The miners already had to show their identification cards before entering the shafts and were forbidden to use the word *uranium* in private conversation (the correct term was *ore* or *metal*). They were now searched for smuggled radioactive material at shift change; guards were assigned to pass the wand of a Geiger counter over their bodies. Suspicion fell on all ranks. Russian supervisors were usually rotated back to Moscow after only a short tour of duty in the uranium zone, to reduce the chances of their making friendly contact with the West.

Suspicion of espionage meant death. Miners, too, were punished with a firing squad for sedition.

Certain that Russian uranium reserves were scarce, America was caught off guard by the August 29, 1949, test of the first Soviet atomic bomb, nicknamed "Joe-1." More than 50 percent of the uranium for that weapon was believed to have come from Wismut. The CIA's forecast of Soviet nuclear potency was off by four years; the embarrassed head of the scientific division termed it "an almost total failure." But the agency was more successful in discovering an enrichment plant near the city of Tomsk after a science officer named John R. Craig was given a furry hat worn by a German-speaking defector who had lived in the area. The hat revealed traces of U-235 in its fur: The uranium was literally floating in the air outside the poorly contained facility. Another informant also helped identify a plant near the tracks of the Trans-Siberian Railroad that bore a misleading sign outside that read STALIN MOTOR WORKS.

Rumors of Wismut's true purpose had been spreading all over the Soviet-controlled zones. The worst stories were fiction, but the place still inspired dread. A West German named Werner Knop acquired forged traveling papers and spent a week at the edge of the mining zone, in the grim city of Chemnitz, called the Gate of Tears by the trainloads of workers who passed through there on the way to the mines, about to suffer what they believed to be displaced Russian vengeance for Hitler's decision to invade the Soviet Union.

"The uranium miners work up to twelve hours a day, urged on by Soviet convict soldiers, who act as overseers, and who are themselves punished draconically if their charges fail to meet the daily norms," wrote Knop in a 1949 book entitled *Prowling Russia's Forbidden Zone: A Secret Journey into Soviet Germany.* "There is no mechanical help, no ventilation, and the most elementary safety devices and health precautions are lacking. The miners work knee-deep in water and are exposed to radioactivity. To this come the ravages of syphilis spread through the brothels established by the Russians for both troops and miners and supplied with the dregs of the big cities of the zone."

Knop's account suffered from exaggeration, but his report of sexual license at Wismut was on target. Women recruited or sentenced to the camps—there were as many as twenty thousand of these "ore angels," as they were known—were treated as playthings by their Soviet overseers and German coworkers. Some prostituted simply to eat. More than half of the workforce suffered from syphilis and gonorrhea.

Knop was also right about the sorry state of repair in the mines. The poor equipment was the direct cause of multiple accidents, including a particularly gruesome incident in April 1947 when an elevator cable broke in mine number 3 and sent eight men plunging to their deaths four hundred feet below. An internal report that year was blunt: "The conditions of work are pictured in such a way that one might believe these are reports from a penal colony. On the surface or underground, the people stand knee-deep in slime, without rubber boots or water boots." Malnutrition was also a factor, as the managers slashed food rations when quotas were not met.

Another report revealed that 1,281 miners had been killed in accidents, and approximately 20,000 had suffered injuries or unspecified damage to their health during a six-month period. To put that figure in perspective, this happened near the peak of Wismut's production, at the height of the Soviet crash atomic program. There were nearly 150,000 laborers in the uranium fields, a number that equals the present-day population of Salem, Oregon. If the company's own figures are to be believed, nearly one out of every seven people wound up dead, sick, or hurt.

Collapses and fires were the most common hazards in the web of tunnels that had been dug and secured by novice engineers. The stone ceilings were known to buckle and cave in if they were not securely braced with logs or metal beams. The Russians often blamed mine accidents on "sabotage" and used them as excuses to weed the ranks of workers they did not trust.

A sign hanging in one of the shafts instructed miners to tap out messages to their rescuers if they became trapped behind a rockfall:

1X = I AM ALL RIGHT.

2X = AIR IS RUNNING OUT.

3X = MY CONDITION IS BAD.

4X = EXTRAORDINARY MEASURES ARE NECESSARY.

One of the deadliest fires broke out under Schlema on the night of July 15, 1955. A worker named Wolfgang Abenroth at the end of a cul-de-sac tunnel did not hear the alarm and was wondering why the pit boss was late for his usual visit. But when Abenroth and his crew smelled smoke, they knew they were in trouble. They stacked up logs and waste rock to build an improvised firewall and sat down to wait for rescue.

"The fear was as big as anything we felt during the war," Abenroth recalled later. "Being walled in for good was always going through my head." Rescuers took more than two days to reach him and his friends. When they emerged into sunlight, they learned that thirty-three of their comrades had been killed. The cause was a frayed ventilator wire in a shaft, but the managers announced it had been an "act of sabotage" and sent in a special detachment of police to investigate.

Abenroth felt guilty for surviving and never forgot the claustrophobia of those two days. "There is a feeling of luck and sadness that will always accompany me like a shadow," he told a newspaper reporter forty years later.

This was an uncharacteristic display of emotion for a miner. The Wismut men were supposed to maintain their composure in the face of danger. To show fright was to risk a beating from the overseer, as well as the contempt of the rest of their brigade. Sangfroid was a prized attribute.

"It was the same thing for the sailors who went into the Bermuda Triangle," said a pit boss named Rudolf Dietel years later. "They knew about the danger, but they sailed in there anyhow. If you let yourself get scared, you weren't much of a miner."

He remembered going to the miners' union beer hall for a funeral wake for a man who had been crushed to death in a subterranean accident. His widow was in a corner booth, crying and inconsolable. "Why did my husband have to work there?" she asked.

Dietel had no answer for her. It was a meaningless question in any case. Death was a routine matter at Wismut; all who lowered themselves into the shafts had to be prepared for it. Uranium mining was the only steady work available, and the only source of income most of them had ever known. This was the Ore Mountains, the country of iron courage and *glück auf*, and uranium was their new silver, whether they liked it or not.

"We all had to do it," Dietel said. "We had no food before uranium. It was a piece of bread and two potatoes to me."

In America, the hype mounted. By 1955, *Life* magazine was reporting—inaccurately—that "more man-hours have been spent in the quest for uranium than were spent seeking all the other metals in history." The following year, *True West* magazine announced, "The most fabulous buried treasure of all time lies scattered and unclaimed, free for the taking."

The get-rich-quick spirit of uranium found its way onto a popular board game, Milton Bradley's The Game of Life, in which players moved automobiles down a twisting path that represented the journey from young adulthood to old age. One coveted square read DISCOVER URANIUM! COLLECT $240,000. Another board game, Uranium Rush, included a battery-powered "Geiger counter" that buzzed whenever a player found the mineral. Bogus "Uranium Clinics" opened in cheap storefronts, promising that exposure to the raw ore could cure arthritis.

In Salt Lake City, one of the most conservative large cities in America, a fraud-laden penny stock market was thriving. A promoter named Jay Walters had purchased a few long-shot claims near Moab and sold shares of his new "Uranium Oil and Trading Company" for 1 cent over the counter of a downtown coffee shop. The stock ballooned more than 500 percent within the month. Permissive state securities laws allowed new companies to sell shares without making any filing with the federal Securities and Exchange Commission, and any unproven company that put *uranium* in the title was bound to appreciate. Venture capitalists were eager to buy any claims at all, so long as they were near Moab; the ghostly moonscape had suddenly become some of the most sought-after real estate in the country. Jerry Anderson's father, who hunted for uranium when he wasn't ranching cattle, was besieged with telephone calls. "At one time I think he had on his desk offers from well over a hundred people who would buy anything he could stake with any count at all," recalled Anderson. "If it just flickered the needle on the Geiger counter, that's all."

There was something about the metal that seemed to inspire an irrational optimism. "Who are buying? Housewives, doctors, big and little businessmen, teachers, bank officers, cab drivers, people in a wide range of economic circumstances," wrote Jack R. Ryan of the *New York Times* in 1954. "Most are spurred by the same impulse that sends millions to race tracks to bet on the horses—the hope for an easy dollar. But stock promoters say there also seems to be a romantic attraction about uranium that makes its mining shares easy to sell." Ryan went on to quote a silver-tongued promoter: "These people can picture 'their' miners hacking the ore of atomic fuel from the rugged mesas and feel they're part of a conquest as colorful as any gold rush—and a lot more important."

There was also a patriotic angle to the fervor: The buyers could justify their purchases almost like war bonds, with the thought that the United States needed to build up an arsenal to fight Communism. The number of penny stock traders in Salt Lake City went from 20 to 112 within a year, and the city became known, somewhat derisively, as the Wall Street of Uranium. But no Mi Vida–size strikes materialized to justify the inflated prices. Stocks almost invariably cratered, leaving buyers with file folders full of worthless certificates.

The boom in Utah inspired a televised episode of the cartoon *Popeye*, titled "Uranium on the Cranium," in which Popeye and Olive Oyl discover radioactive treasure on a desert island. They are attacked by their greedy nemesis, Brutus, who is disguised in an ape suit. Then a real ape knocks Popeye high into a tree. The cartoon ends on a happy note: Popeye eats a can of spinach, decks both Brutus and the ape, and proceeds to mine the uranium himself, Olive Oyl at his side.

A more dubious Hollywood effort was *Dig That Uranium*, a 1956 vehicle for the aging Bowery Boys, who buy the deed to a uranium mine and head to Panther Pass, Nevada (actually, Iverson Movie Ranch in California's San Fernando Valley), to get rich. The uranium craze was an excuse to place the comedians in a Western setting, where all manner of cowboy stereotypes could be goofed upon: poker games, gunfights, campfires, and stoic Indians who wrap themselves in blankets and stare into space. The film is also noteworthy for a cameo by Carl "Alfalfa" Switzer, playing a huckster who sells the boys the worthless* property.

Two years after the movie's release, Charlie Steen started campaigning for a seat in the Utah senate. Although a determined atheist, he was still popular in his heavily Mormon district for his image of crusading atomic patriotism in the fight against Communism and he appeared to be on the verge of unseating the incumbent. One local radio reporter characterized the race as "George Hurst, the Mormon, versus Charlie Steen, the American."

Shortly before the election, Steen took a short break and flew his plane down to El Paso, Texas, to accept an outstanding-alumnus award from the Ex-Student Association of his alma mater, the Texas College of Mines. He had been a profligate donor to the college ever since his lucky strike and, during his occasional visits to campus, had been in the habit of taking geology students over the Mexican border to Ciudad Juárez for rounds of tequila shots. He was irritated, however, that the school had changed its name to Texas Western College in an attempt to appeal to a broader range of students. He was also unhappy to see that his college was offering a new range of liberal arts and vocational courses, many of which had nothing to do with mining.

When he rose to say a few words at the banquet, there was applause. But this would not be typical after-dinner oatmeal. Perhaps Steen was worn down by the grind of his Senate campaign, tired of having to shake hands and say careful things in front of the voters. Perhaps he had been rankled by the constant lawsuits from his former business partners and competitors. For whatever reason, the discoverer of Mi Vida was in a vinegary mood that night and, in words documented by the journalist Raye Ringholz, he let his college friends have the brunt of it.

"I know the proper way to accept this award," Steen began. "I am expected to say 'thank you' and sit down. However, inasmuch as I did not seek this award, and as Dean Thomas reminded me last night that I was

---

* This was the only realistic element of the movie.

the only son of a bitch he knew who had made a career at being one, and was a success as a result, you need not expect the proper response."

Steen then began reading—contemptuously—from the school's course catalog. He cited a list of classes he considered frivolous. They included Coaching Basketball, Real Estate Brokerage, and Baton Twirling. He suggested the list be expanded to include "Beer Guzzling, a course in how to chug-a-lug a gallon pitcher without getting a permanent crease on the bridge of your nose" and "Mexican Relations, how to go to Juárez and keep enough money to get back across the bridge." Steen also insisted he would not accept his award unless the name engraved upon it was corrected to read TEXAS COLLEGE OF MINES and not the name of an institution he did not recognize. With the audience now coughing and squirming, he reached his peroration.

"This son of a bitch previously mentioned, at thirty-nine years of age, is a living legend of the uranium boom that he helped create, a boom that raised the U.S.A. from a 'have-not' nation to the number-one position of uranium reserves in the world. Whether he dies a multimillionaire or a broken down, ragged-ass prospecting tramp, his place in the mining history of our country is secure."

When he sat down, there was embarrassed silence and a few boos. The diatribe prompted Paul H. Carleton, the president of the alumni association, to write an open letter to Texas Western's administration the next morning. He apologized for not rising to stop Steen's "abuse and blasphemy" and regretted that the tone of the event was not focused more upon "living in a peaceful, understanding world."

Charlie Steen apologized for none of his words and snapped off an angry reply to Carleton.

"As for 'living with our fellow men in a peaceful, understanding world,' what sandpile have you buried your head in since you got out of college?" Steen wrote. "We are living in a world in which two systems are locked in mortal conflict to determine which kind of world our kids are going to inherit. . . ."

In East Germany, the lumps of uranium had a nickname among the freight handlers. They called it *Heilerde*, which means "holy ground."

The name was partly ironic—they suspected the hazards of the black ore they were loading onto the railroad cars—but there was an earnest aspect to the name as well. The uranium was a reason they had a paying job and was what separated them from the lower ranks of society and those still suffering from the deprivations and hunger caused by the war.

The underground brigades had their own nickname for the veins of pitchblende they were trying to intersect. They called it *Speck*, or "bacon." The name was appropriate, because some fundamental changes started to take place in the German uranium fields in the years after Stalin's death in 1953.

The Red Army stopped its direct oversight of the mines, dismissed Maltsev, and passed control to a reconstituted state agency called SDAG. East German authorities now became shareholders in the new corporation and had more direct influence in the day-to-day operations of the network of mines, mills, and railways. The population of laborers also leveled out to a steady forty-five thousand as prisoners were allowed to go free. The new directors were eager to shed their image as cruel taskmasters and reevaluated their methods of encouraging productivity. More attention was paid to worker safety, which boosted the morale of the drilling crews. But most important, Wismut started emphasizing a new principle: namely, uranium can make you rich.

Personal gain had been an inducement at Wismut from the start, albeit in a more muted fashion. Those who had volunteered for mine work in the 1940s were issued daily "Stalin parcels" of soup, coffee, bread, and cheese above the portions granted to other German citizens. They had always eaten better than their neighbors. But now a Wismut man could earn more hard currency than almost any other class in the Socialist state. With bonuses, a miner could expect to bring home 3,000 marks a month, which compared favorably with the average salary for a village mayor, who made 250 marks per month. Refrigerators and appliances and other luxuries could be acquired much more quickly here than anywhere else in East Germany. It might take only two years to acquire a car, whereas others in less exciting professions had to wait up to seventeen years.

To acquire this soft life, one had to dig out more pitchblende than the quotas required. The brigades were thus set up to compete against one another, with tantalizing payoffs for the winners and collective scorn for the losers. The uranium had been transformed into *Speck*—something to be pursued instead of having one's nose forced into it.

This was an obvious departure from Communist ideology. But it represented an important psychological insight. Machiavelli once said it was better for a king to be feared than loved. But greed is an even better motivator than fear. It is more sustainable over the long run and is, in fact, potentially endless. And it cuts down on laziness.

Walter Hegenbart was one of those attracted by the high wages. He signed up as a teenager after he saw his brother open a wallet stuffed full of marks after he spent only a few weeks on a drilling brigade. This was a powerful inducement: Hegenbart and his brother and their eleven siblings had been forced out of their home in the Sudetenland in 1945 and had no other way to earn money. "We did it because we were hungry," he explained later.

Hegenbart was assigned to a six-man brigade, and the men learned to depend on one another. Those who appeared to be slacking were subject to harassment and abuse, for it put the entire unit in danger of losing a bonus and looking incompetent in front of the other brigades. The biggest payouts were for pitchblende—the *Speck*—weighed by the kilogram, but additional money was offered for other good behavior, such as time worked without an accident and perfect attendance and taking good care of tools. Penalties could be assessed for following incorrect safety procedures, but pit bosses tended to look the other way as long as the uranium kept coming.

Taking a break during the twelve-hour shift was unthinkable. Many skipped lunch, preferring to drill at the walls instead. And going above-ground to use the toilet was a sure ticket to ostracism, so most defecated inside metal buckets, known as *Toilettenwannen*. Though the hard-rock miners were legendary for their love of brandy and beer, few drank it underground, where it could hamper efficiency and get them behind on their bonuses.

Temperance faltered, however, at week's end when the bonuses arrived. The bars in Schlema were periodically told to close down on payday: There was bingeing and brawls, and some wives complained they never saw the money before it was spent. The pay and the generous rations were lucrative enough to attract occasional defectors from West Germany, such as a man named Herbert Kampf, who was held up as a model convert to the Socialist cause. He obligingly told a local newspaper, "I follow my brigade."

There were financial pressures underground, but there were also bonds of loyalty. The men gave playful nicknames to one another. If a miner chafed against his name, it would be his to wear forever. One particularly slow worker was called the Master Driller. An officious pit boss with a mustache was called Fake Hitler. A party lackey was known as Lenin. The brigades were issued shoulder patches as Boy Scout troops were, and they marched in dress uniforms in military-like formation on the days when there were parades. Many asked to be buried in their Wismut uniform when they died, including even Heinz Pickert, the barber whose hands had been ruined.

Every New Year's Day, the men gathered at the miners' union beer hall, a log-walled place called Die Aktivist, where a list of names was read: all those who had been killed in accidents the previous year. Each name was followed by a single chime of the elevator bell mounted on the wall. On other nights, the liquor and the boasting would flow until midnight. Alcohol had always been one of the perks of Wismut—the standard ration of brandy was ten bottles a month, and it was said to be an effective tonic for the mysterious coughing spells that many were experiencing. The booze received the nickname "Miner's Death."

In their cups, the men would sometimes sing the old songs of the Ore Mountains. Few of them had been born in the hills—most had been dragged there by force—but they saw themselves in the lineage of those proud men who tore silver from the earth and got paid for it.

> Luck up! Luck up!
> Here comes the pit boss
> And he has his bright light
> In the night
> Already lit!

Wismut had become, by that point, a "state within a state," in the words of the historian Rainer Karlsch, functioning as a semiautonomous fiefdom within East Germany. There were seventeen towns under its watch. It had its own hospitals, police, and court system. The mines had contributed up to 80 percent of the uranium for the Soviet nuclear program, and its managers were rewarded with generous allocations of equipment.

The men and women of Wismut were reminded of their duty to Communism, and how the uranium played an important role in stopping the global menace of capitalism. They were told that America had been threatening the USSR with nuclear weapons and that only a strong counterarsenal could save lives. "Everyone believed that since they had bombed Nagasaki, [the Americans] would bomb us, too," recalled Walter Hegenbart. Signs all over proclaimed ORE FOR PEACE.

The first promotional film for Wismut was shot in 1959. Laced with strange homoerotic imagery, it depicts a team of three men stripped down to their shorts, rhythmically unloading piles of uranium ore from cars to the sound of orchestra music. Others are pushing blasting caps into walls dripping with moisture, and another crew is hosing debris off the cart tracks. Tulip-shaped buckets dump the black stones in the crushing mill. The uranium miner is portrayed as a virile national hero.

A grittier view was presented in the movie *Sunseekers*, directed by the auteur Konrad Wolf, who happened to be the brother of the national spymaster Markus Wolf. The director Wolf was regarded as a creative genius in his native Germany and he had credibility with the Communist authorities. In 1957, Wolf received permission to shoot a new film in the heart of the uranium fields.

In retrospect, it is a miracle that *Sunseekers* was ever made. It tells the story of a young woman named Lutz who is arrested in a barroom brawl and sentenced to mine uranium in the Ore Mountains, depicted here as a grim wasteland of rock heaps and muddy roads, with a savage mix of convicts, ex–SS soldiers, and Communists forced to live there in quest of radioactive fuel. The mostly German miners are lorded over by their Russian overseers, who demand "uranium every hour" and are slow to replace fraying electric cables in the mines. Lutz at first falls in love with an alcoholic miner named Gunther, who insists, "This is the best work there is! You can hear the uranium crackling!" But she eventually winds up with a moody one-armed pit boss named Beier, who seems to embody all the sufferings of postwar Germany. He dies underground in a fire started by one of the defective cables—literally trapped in a tomb of uranium.

The film is not without some patriotic content. The opening crawl reads: "The flash over Japan was meant to illuminate the American century. To protect itself and help world peace, the USSR had to break the atomic bomb monopoly." But taken as a whole, the movie is a strong critique of the uranium-hungry policies that refashioned this corner of Germany into a zone of despair. The mines are depicted as dangerous and dirty; the landscape itself is without cheer or hope. The title *Sunseekers* is also a sly piece of wordplay. The men and women who went to work for Wismut were seeking not just the mineral whose power was equivalent to the sun but also some warmth in their own lives. The miners who spent half of their day in darkness were also eager for the return of sunlight that accompanied the end of each shift when the elevator brought them aboveground again. This was a journey that, from the deepest shafts, could take up to two hours.

Filming on *Sunseekers* was completed in 1959, and the movie was scheduled to premiere in the town of Schlema, at the very heart of uranium country. Engraved invitations were sent out to Communist Party leaders and ordinary miners. But days before the debut, it was reviewed once more by Soviet censors, who objected to the dismal portrayal of life inside and outside the mines, and especially the insinuation that Russian supervisors were indifferent to the safety of their German comrades. The showing was canceled, and the movie was put on the shelf for twelve years before a more relaxed East German regime permitted its release in 1972.

This would not be the only unflattering portrayal that party officials tried to suppress. In 1988, a sensational book-length study called *Pechblende* began creating a stir among reporters and academics in West Germany. It had already been circulating in the East through *samizdat*— that old Communist method of quietly passing copies of banned material from reader to reader. The author was a twenty-four-year-old university student named Michael Beleites, who had surveyed the human and environmental damage in the Ore Mountains. Beleites published his findings under the cover of the Lutheran Research Institute, which was out of the reach of party censors, but he suffered harassment in the aftermath. His mail was opened, his parents' telephone was tapped, and he was followed by the Stasi secret police whenever he left his apartment. Beleites later said the only thing that saved him from prison or execution was fear among Stasi officers that his disappearance would only throw more light on the environmental disaster at Wismut.

Pechblende was written in a calm but relentless voice. It told a grim tale of environmental wreckage in the closed uranium towns, and of people living there seemingly ignorant of the health risks of close contact with what was still euphoniously called ore. Mushrooms growing on some of the toxic waste heaps were frequently picked for salads and cooking. Some of the poured concrete in the newer buildings was made of ground-up radioactive rubble. Radon gas was floating in elevated quantities all over the Ore Mountains.

The report included a letter from the daughter of a middle-aged miner that put words to one of the realities about uranium mining that many suspected, but few wanted to acknowledge out loud.

"I can't forget how he sat there in the bare hospital room, alone, perplexed, and hopeless. He just kept shaking his head. All he wanted was to make his own decision when to die. He had seen too many other friends die. My father jumped out of the window, eleventh floor. He knew nothing, nothing about radiation. Right near the end a doctor finally told my mother it was lung cancer."

Uranium is always disintegrating. This is the signature trait that distinguishes it from most of the elements in the periodic table and what gives it its power. During the Jurassic period, when it came to rest inside the dinosaur corpses trapped in the sandstone in America and also mingled with the silver-bearing deposits of Germany, it remained unstable at the core, continuing its spiral downward through the decay sequence. One of the first things spawned is radium, the element prized by Pierre and Marie Curie. Radium, in turn, decays into radon-222, the heaviest known gas in the natural world. It is called a noble gas because, like neon and helium, it cannot bond with any other element. Unlike the other nobles, however, it is constantly throwing off pieces of itself and becoming, yet again, something new. It has a half-life of just under four days.

Air pockets in the uranium ore house the entirety of the radon decay sequence: from radium-226 to radon-222 to polonium-218 to lead-214 to bismuth-214, and so on until everything comes finally to rest at lead-206. Down the isotopic chain, from instability to instability, a series of "radon daughter" elements are released, radioactive particles with half-lives ranging from twenty-seven minutes to a fraction of a second. They are tasteless, odorless, colorless, and invisible. If these gaseous particles are locked inside the ore, they are harmless. But the uranium inside the Colorado Plateau and the Ore Mountains had settled into porous formations of sandstone and clay. These rocks have countless tiny airholes through which the gas can seep outward. The greater the surface area exposed, the more gas escapes. A long flat surface like the wall of a mine tunnel provides an ideal environment for radioactive gas to respire from the surface. And when the conglomerate ore is chipped or crushed, even more radon daughters escape from spongelike chambers within the rock.

This effect is relatively harmless in outdoor spaces because the radon daughters almost instantly dilute into an ocean of fresh air. But inside a confined space such as a crushing mill or an underground mine, the radon daughters fasten themselves to motes suspended in the air: dust, particulate, and water droplets. These can be breathed deep inside the lungs. Most of them are harmlessly exhaled. But a few remain trapped in the soft pink tissue and alveoli that form the interior surface of the lungs. These pieces of radioactivity then literally become part of the body, often lodging into bone marrow. The body absorbs them like calcium, and the particles keep irradiating the body from the inside. They fire their protons and neutrons into neighboring cells, ripping through them like machine-gun bullets. The cells struggle to repair themselves, and a few mutations are eventually created. These mutations replicate and eventually become cancer. A day-after-day exposure to radon dust poses a risk for lung cancer that is four times above the average. The usual incubation period is fifteen years.

The U.S. Atomic Energy Commission knew of these dangers even before Hiroshima. An agency health official based in Grand Junction, Colorado, named Ralph Batie had read about the mysterious "mountain sickness" of the radium miners in the Ore Mountains. He also knew of the high cancer rates in some of the New Jersey factories where workers, mostly female, had painted luminescent dials for submarines and aircraft during the buildup to World War II. Many of the women were in the habit of licking their paintbrushes to straighten the bristles; their teeth fell out as a result. Some were also in the habit of dabbing their lips with the glow-in-the-dark paint for a showstopping effect in dark nightclubs.

Growing suspicious, Ralph Batie took air samples in several mines on the Colorado Plateau in the late 1940s and found alarmingly high concentrations of radon gas and other alpha ray–emitting airborne material. These were at levels thousands of times higher than the AEC's own regulations would have permitted in an enrichment plant or a bomb assembly factory. Furthermore, the uranium miners often practiced "dry drilling," which meant hammering away at a wall with a pneumatic hammer and no lubricating water to keep down the dust, which was loaded with alpha-emitting radon. After these findings were forwarded to Washington, the AEC did an about-face and said it was outside their jurisdiction to tell a private mining enterprise how to regulate its air quality. That, they said, was the responsibility of the state health agencies of Colorado, Arizona, and Utah. But those authorities lacked the staff or the political muscle to accomplish anything meaningful.

"They weren't getting paid much and they weren't very diligent," recalled the miner John Black. "The mine inspector was looking to see if there were some loose rocks or if you were using a short fuse. . . . Ventilation was the last thing on their minds. By the time they showed up again, we might be thirty miles away in another mine."

The inertia continued for almost twenty years. Batie was hounded out of his job and forced to transfer to a different office. Other officials who tried to sound an alarm were also treated like pariahs and troublemakers. When Duncan Holaday, a radiation expert with the U.S. Public Health Service, warned in a 1952 report that air samples taken from Utah uranium mines contained deadly amounts of radon gas, he was told that he should continue to "study" the problem on a long-term basis and make no public statements. No action was taken, even though the evidence was damning and almost painfully obvious. Holaday had reported seeing a "yellow coating on tongue and teeth" among the miners. Holaday's report was not widely circulated, and his work was carefully edited by AEC officials thereafter. Going public was not a realistic option; challenging the government's national security prerogatives would have been viewed as treachery at that point in the cold war. The only worry about radioactivity was that negative publicity might slow down production.

"There is no doubt that we are faced with a problem which, if not handled properly, could be made public," wrote the AEC official Jesse Johnson in a 1952 memo. Widespread awareness of dangerous mine conditions "could adversely affect uranium production in this country and abroad." More alarming for Johnson: "Communist propagandists may utilize any sensational statements or news reports to hamper or restrict uranium production in foreign fields, particularly at Shinkolobwe."

This attitude was shared by many in the Southwestern uranium fields, where the workers saw themselves as soldiers of the cold war. "Finding and processing fuel for atomic bombs was the patriotic thing to do," wrote Tom McCourt, who lived near a leaking and dangerous mill in the town of White Canyon, Utah. "The Soviet menace could be kept in check only by recruiting the help of the Atomic Monster. It was a pact with the Devil, but considered to be worth the risks."

The first lung cancers appeared five years after the AEC's bonus program was announced. By 1966, nearly one hundred miners were dead. In the face of overwhelming evidence, and with families beginning to protest, the U.S. Department of Labor was persuaded to force mines to maintain air quality with a fixed limit on radon. Companies were required either to drill airholes from above or to blow radioactive dust out with electric fans. Some companies protested the rule as being cost prohibitive, and one even tried to make the novel claim that the blowing dust particles would create eye injuries.

The retrofitting turned out to be cheap, though, as production had slowed to a crawl. The U.S. military had stockpiled more than thirty thousand strategic nuclear warheads, enough to obliterate most of the cities in the Soviet Union and its satellite states many times over. Over-supply of uranium had become a serious problem. In the mid-1960s, the government phased out the buying policies that had encouraged so many prospectors to roam the deserts. The latest mineral rush in the United States had come to a close.

The AEC retained its status as the sole buyer of domestic uranium with an eye toward serving nuclear power plants. Most of the shallow deposits had already been found and exploited. Deep-drilling equipment was needed to get at what was left, and well-established companies such as Sohio, Exxon, Union Carbide, Kerr-McGee, and Getty Oil were largely dominating the game. With deep budgets, they installed fans to blow out dusty mines.

But it was too late for those who had worked at the peak of the boom. The mortality rate kept creeping upward. Especially hard hit was the Navajo Indian reservation, where lung cancer was a rarity until uranium came along.

The Navajo are the largest Indian tribe in America, occupying a res-ervation covering an impasto of canyons and mountains in northern Ari-zona, and smaller parts of Utah and New Mexico. They had migrated here from present-day Alaska in the early sixteenth century, about the same period in history when the first silver trenches at St. Joachimsthal were being dug. The Navajo call themselves Diné, which means "the People," and many keep a ceremonial dome of mud and wood called a hogan; its small door always faces east toward dawn. Sheep are a staple food and a primary measurement of wealth; blanket weaving from sheep's wool is a signature craft. Navajo country is home to four mountain peaks considered sacred, and also a substantial amount of uranium. The bright yellow surface ore was a common additive to religious-themed sandpaint-ings. When the race to build an atomic bomb geared up in the 1940s, this became a hot place to mine. Outsiders could not obtain a mining lease, but companies typically skirted this by recruiting a Navajo "foreman" (ideally with political connections on the tribal council) to get the license and enlist a few dozen friends and relatives to do the labor inside the sweltering dog holes. Drinking water was sometimes from runoff tainted by uranium dust. Workers were also in the habit of sucking water from the trickles on the sandstone walls. Pay was often as low as a dollar per day, and the foreman would sell the ore to the sponsoring company for a 2 percent royalty. It was an exploitative arrangement, but jobs were scarce on the dirt-poor reservation and the patriotic fervor surrounding uranium made it a hard bargain to refuse. By the end of the boom, the Navajo cancer rate had doubled, and the land was scarred with more than thirteen hundred abandoned mines, as well as with slag heaps that still emit clouds of gaseous radon. Kept in the dark about radioactivity, some Navajo used the tailings to make concrete for their houses. Others kept their sheep penned in the crevasses of mines, then ate their radioactive meat.

One of the many sickened was Willie Johnson, a Navajo who worked summers at a small uranium mine to put himself through school. "We'd dig out the uranium with a shovel and a pick, dump it in the car, and send it out," he told an interviewer in 1998. "No one told us that mining was dangerous. We wore hard hats and steel-toed shoes, no protective clothing. There was never enough ventilation down there, and lots of dust." Said Ben Jones, another Navajo miner, "When you blew your nose, it was yellow dust." A doctor would later testify that working in a mine such as this was the equivalent of receiving a daily chest X-ray for six months straight.

At least six hundred Americans—Navajo and white alike—would eventually die from illnesses linked to the radioactive dust. After a lengthy lobbying effort from widows and family members, Congress passed the 1990 Radiation Exposure Compensation Act, which granted $100,000 "compassion payments" to miners diagnosed with cancer or other respiratory ailments. The law was amended ten years later to allow payments for mill workers as well. Part of the argument was that miners and mill employees had been working in the name of war, building up the American arsenal, so it was proper that they should be treated as if they were veterans. A U.S. Public Health Service physician would admit that the government had "largely ignored" the risks of radon gas. Former interior secretary Stewart Udall sued on behalf of cancer-stricken Navajo miners, but lost after a federal judge ruled that the AEC had been acting in the name of "national security."*

The death toll was far greater in the Ore Mountains, where the work shifts were longer and many more people were employed over the years. Dry drilling was halted and ventilation equipment was installed in 1970, on about the same schedule as the American retrofitting, but the health damage was widespread and permanent. The reunified German government ended up paying compensation to 7,695 miners who developed lung cancer. A British medical study later found that the average lung cancer risk for employees of Wismut, in jobs both in and out of the shafts—miners, clerks, supervisors, cooks, and guards—was slightly over 4 percent. There is no authoritative death count, but it has been estimated at 16,000.

Wismut officially went out of the uranium business when the German Democratic Republic collapsed in 1990. The remnants of the company were quickly reconstituted as a quasistate agency called Wismut GmbH and given the task of cleaning up the massive environmental damage its predecessor had created. Dozens of open pits had to be filled in. Many required filler in addition to their tailings because of the effect (well known to grave diggers) of soil compacting as it is repacked. Underground shafts had to be sealed with concrete, and ponds had to be drained of radioactive water. Those ore piles not pushed back into pits were covered with grass and trees. One pit is now a golf course. The entire project is expected to cost the equivalent of $8 billion, almost all of it coming from the federal government.

"We have to make the countryside livable again," a Wismut engineer told me. "The most important part is to give it back to the public."

The valley ground where Schlema's central district used to stand has dropped about forty feet lower than its elevation before World War II. The web of uranium tunnels underneath the city created a subsidence effect. There is now a giant grassy park where the shafts and barracks and ore

---

* Uranium mining is forbidden today in the Navajo Nation, despite the reservation's abundant reserves: the "Saudi Arabia of uranium," one frustrated company official described it to a reporter.

piles marked the heart of Germany's uranium country. A sculpture of a wind sail now sits at the middle of the grassy area, which is called a health park. Granite slabs mark the places where each headframe once pulled the innards of the earth onto the surface. The city's new motto is "Natural Schlema," and a handful of new resorts have opened for business, hoping to recapture some of the Weimar-era fad associated with the dubious health benefits from drinking and bathing in radium-heated water.

I went to see Michael Beleites, the environmental researcher whose *samizdat* report on Wismut had embarrassed the authorities and nearly got him thrown in prison two decades ago. He now works for the government of the state of Saxony as curator of declassified files from the Stasi. Beleites wore a navy blue sweater underneath a tweed jacket. He had the bearing of a modish professor of Elizabethan literature at a small liberal arts college. On his desk in his office in Dresden were a laptop computer and a metal lamp with a curved neck as slim as a crane's. He told me he felt that southeast Germany was still refusing to face its heritage as a uranium colony.

"The story most people believe around here is that Stalin had to break the West's atomic monopoly and that Germany helped secure the peace," he said. "I see it differently. It did not secure peace. That uranium helped ensure a continuation of more and more oppression. Even though it was never used in war, it created fear and dread all over the world. And we got nothing except these huge rehabilitation projects and a lot of dead miners. 'Ore for peace'? What a stupid idea!"

On the other side of the mountains in St. Joachimsthal, the marble-sheathed Radium Palace Hotel reopened in the sixties as a getaway for Communist Party officials who hoped that a good soak would bring them relief from their arthritis and gout. In 1981, the French atomic scientist Bernard Goldschmidt managed to secure a visa and paid a visit to St. Joachimsthal, where so many people had been forced to mine uranium for the Soviet nuclear program. The camps had closed down several years before, and the mines were operating at a fraction of their capacity, but Goldschmidt had still been curious to see the birthplace of the mineral that had so profoundly affected the course of world history.

He recalled: "My pilgrimage to the Saint-Joachim Valley, with its evocative memories of its successive masters, counts, kings, emperors, presidents, Fuhrer, and party general secretaries, of its generations of miners subjected to dangerous work, of its empires of silver, radium, and uranium, thus ended with a vision of a few fat gentlemen, full of illusions and hope, doing exercises stark naked in a swimming pool of radioactive water."

In the winter of 2007, I paid my own pilgrimage to St. Joachimsthal. The Radium Palace Hotel is still in business and now caters to elderly tourists seeking relief from arthritis or rheumatism. The hotel's promotional brochures are printed in a variety of languages, including Arabic. Face-lifts, acupuncture, and Botox injections are offered as supplements to the hot radium baths. The place was almost deserted the night I stayed there. Its yawning ceilings, empty corridors, and chessboard floors were reminiscent of a royal mausoleum, or a set for a remake of *The Shining*.

The next morning, I hiked up through the village and past the St. Joachim church to the head house of the Concord mine, the oldest of the sixteenth-century silver shafts, which had since been dug down a third of a mile. Almost all of the uranium is gone now, but there are three springs inside that still pump radioactive water to the Palace and other spa hotels in the village.

The mine shaft is usually not open to the public, but I met with an engineer named Jiri Pihera who spoke a little English. He eventually agreed to take me down the lift to show me the chambers from which uranium had first been brought out to the world and the U.S. dollar had taken its name.

Pihera handed me a lamped helmet and a rubber coat and took me into the elevator cage, which dropped us down to the twelfth and deepest level. The lift opened to a well-lit room full of cylindrical pumps, roaring in monotone fugue. "We have to pump a thousand liters a minute out of here, or else the place would flood," Pihera told me over the racket. "From this basin, we pump the water to a second basin three hundred meters up. And then it flows three kilometers downhill to the spas."

He switched on his miner's lamp and led me down a corridor that had been braced with giant metal bands. It was like walking through the rib cage of a fish. I had to duck in places to avoid banging my head against the supports. Though it had been cold outside and the mountains were frosted with snow, the air in the shaft was warm and torpid. The walls were sweating trickles of groundwater. We climbed a set of rickety wooden stairs and took several turns in the dark corridors. Flakes of gypsum twinkled in the light of Pihera's lamp. After about a quarter mile, we came to a chamber lit with electric bulbs, with a large wooden vat in the middle. A sign in Czech proclaimed this to be a spring called Behounek.

"This is the most important spring we have," Pihera told me. It had been drilled more than forty years ago and turned out 150 gallons of radium-tinged water per minute. He scooped a double palmful of water out of the vat and slurped from it.

"Drink," he suggested. "It's very good for your health. You don't want any?" He looked disappointed. "People pay good money for this, you know."

We walked farther along through the tunnels. Cart tracks long out of service were embedded in the floor and dead electrical cables, wrapped in black rubber, striped the walls. A graffiti scrawl in white paint adorned one mine wall. Pihera translated it for me, somebody's long-ago private joke: HERE WORKED A CUDDLY BEAR.

It would have been impossible to calculate the number of miners who had worked St. Joachimsthal's tunnels and drifts in its nearly five centuries of operation. Most of them had been chasing silver. Uranium had emerged from these wet European shafts as an accident, a sick ebony residue abandoned in the forest, its powers unsuspected. Only when science had focused its lens and seen the violence lurking in its guts was it considered treasure. A trash rock could now push the world.

Pihera slowed his pace as we got closer to the man cage. He trained his light on the west wall and inspected it closely. It took him a few minutes to find what he was looking for.

"Come here and look at this," he told me. "We found this a while ago. There is almost none of this left."

He pointed to a small chip, jet black, embedded in the schist. It appeared to be the tip of a seam hiding inside the wall, which had somehow eluded being dug out and carted off.

"The last piece of uranium in the Jachymov mining district," he announced with a smile, using the Czech name for the town.

I ran my thumb over the exposed chip. It was thin, about an inch tall; the size of the edge of a U.S. quarter.

Charlie Steen's mansion on the hill is now a steak restaurant called the Sunset Grill. A sign out front reads MILLION DOLLAR VIEW! DINNERS FROM $13.95. The jeep he was driving when he made the discovery at Mi Vida is parked out of sight next to a side building, its interior rotting. The broken drill rig that made the "million dollar lick" is rusting next to a cliff. His uranium mill closed down in 1984, and the U.S. Department of Energy is still trying to decide how best to contain the radioactive wastes that may be leaking into the nearby Colorado River. The current plan is to load the toxic gravel onto railroad cars for burial at a site thirty miles away, a ten-year project estimated to cost $400 million, which is about a quarter as much as the value of all the uranium produced in the United States through 1964 when adjusted for inflation.

Thousands of uranium mine shafts pockmark the desert, and the rangers at Canyonlands and Arches national parks warn visitors to avoid them. The entrance to the Hidden Splendor mine is littered with boulders; the switchbacking road up the side of the cliff has nearly washed away from fifty seasons of rain. Moab shifted its economy to tourism and stopped calling itself the Uranium Capital of the U.S.A. three decades ago. The access roads blasted into the slickrock are now the delight of four-wheel-drive owners and mountain bikers, who can venture deep into the otherworldly landscape of sandstone spires and fragrant juniper. One of the most striking of the uranium roads is the Shafer Trail, which zigzags nearly fifteen hundred feet from the top of a giant mesa. It had been carved there by six young men from Moab trying to reach an enticing claim. They raised $50,000 from their friends, jackhammered and dynamited their way down the cliff, and built another road halfway up another cliff to the rich Shinarump formation where the Geiger readings had been so promising. They drilled into the cliff and found only sandstone.

"Nobody ever got a pound of ore out of there," Bob Mohler, the only living member of the road crew, told me with a laugh ten years ago. The Shafer Trail is today the most visited man-made attraction in Canyonlands National Park, aside from petroglyphs made on the cliffs by the long-vanished Anasazi Indians.

Nick Murphy, who led the construction team, died in 1996 of Parkinson's disease and not cancer. He never regretted his time searching for uranium. "It has actually been progress," he once told an interviewer. "I can't look at it any other way, because I do think it's the future of your whole damn world if you want to put it that way. It is your next energy source. It is progress. It has got to be."

Vernon Pick, whose tale of finding the Hidden Splendor mine had thrilled the readers of *Life* magazine, developed a paranoia of nuclear war as he grew older. He used his money to build a fortified estate in a remote canyon in British Columbia, where he hoped to be one of the last survivors of an atomic holocaust that he feared was coming. Pick lived as a recluse, but remained enthusiastic about scientific progress and built a hydropower dam to supply electricity to his laboratory before he died in 1986.

Charlie Steen sold his interest in the Mi Vida mine in 1962 and moved to a horse-ranching spread outside Reno, Nevada. A series of bad investments wiped out his fortune, and he was forced to declare bankruptcy after the IRS confiscated his remaining assets for back taxes. The power was shut off in the Reno home. He complained to a reporter, "We're sitting in the middle of all this luxury—my wife, four sons, and I—eating canned beans and stale bread just like we did in that tar-paper shack seventeen years ago before we struck it rich."

Steen decided to go back to the place where he had been happiest—out in the mineral fields, hoping to chance upon the next lode of riches that would turn everything around. In 1971, while drilling for copper in the Deep Springs Valley of California, he was hit on the head by a wrench attached to the drill pipe. It put him in a coma for more than a month, and he woke up with almost no command of language. For a time, the only words he could pronounce were *gold* and *silver*. His recovery was slow, but he never lost hope that he would one day recoup his lost fortune from an old gold-mining claim he had managed to hold on to in the mountains to the west of Boulder, Colorado, a place called the Cash Mine at Gold Hill. "It's not a matter of thinking I'll make another big find. It's *knowing* it," he told the *Salt Lake Tribune*.

His sons began to squabble among themselves for control of the remaining money, as well as the gold claim. Mark and Andrew Steen filed lawsuits against each other and no longer speak. In an unrelated affair, a grandson named Charles Augustus Steen III pleaded guilty to extortion in a San Diego court for purportedly demanding $2.5 million from the widow of Theodor Geisel, the children's author better known as Dr. Seuss. In his correspondence with Audrey Geisel, Steen was said to have mailed one of his paintings, which depicted the Cat in the Hat having sex with a blond woman, who was herself giving oral pleasure to the Grinch. The grandson of the Uranium King received three years probation and was required to complete a course in anger management.

Charlie's wife, Minnie Lee, died in 1997 and left two of her sons a dollar each for what she called "ingratitude and dishonesty." The lawsuits over the Cash Mine and what was left of the uranium fortune created a long and byzantine court battle. "It's not a happy ending, I assure you," said Mark Steen.

"Everyone in this family has spent their lives pursuing this 'pie in the sky,'" said Andrew Steen, to a reporter. "They became obsessed with this nebulous thing that never came to pass. That's what happened to the Steen family, and it's pretty damn tragic."

Charlie Steen, the Uranium King, spent his last years in an Alzheimer's haze. He passed most of his days in front of the television set, not knowing which program was on and not caring. Steen died on New Year's Day, 2006, and his body was cremated, his ashes mingled with his wife's in their favorite silver champagne bucket. Their ashes were scattered together near the mouth of the Mi Vida mine—long since abandoned—which had produced twelve million pounds of uranium ore in the course of its life, enough to make at least eighteen atomic bombs.

6

THE RAINBOW SERPENT

There is a flat-topped mountain in northern Australia that looks like a river barge tipping over into deep water. The sides are cracked and stained tangerine with iron. At the highest point, a cliff juts upward like the prow of a ship and contributes to the illusion that the whole mesa is on the verge of sinking into the forest of paperbark trees below. This cliff is called Mount Brockman, and it is considered a holy site to a local band of Aboriginals, whose religion is concerned, above all, with geography.

They believe the earth was woven together out of the threads of musical notes—specifically, the songs of the Creators, who sang them in a preexistence called the Dreamtime. All the rivers, rocks, deserts, grasslands, and forests of Australia are pieces of frozen music. The songs did not just create the earth, they *are* the earth. The lyrics tell stories about ancestor beings who have the same virtues and failings as humans, similar to the gods of the ancient Greeks, who were also fond of seduction and betrayal. The songs usually have wild animals as characters and are sometimes laced with sex and violence. The most ghoulish details are for only the elders to know. To be trusted with the secret content of myths is a mark of power.

The story associated with Mount Brockman might never have been made public were it not for a geologic quirk of the structure. Thirty years ago, under significant political pressure, the Aboriginals were compelled to explain what had happened in the Dreamtime.

It is this: Mount Brockman is actually named Djidbidjidbi. The orange stains on the walls are not iron, but blood. A pond on the top is the opening to the dwelling place of a king brown snake, sometimes called Dadbe, who had retreated there to sleep. She was a cousin to a greater beast known as the Rainbow Serpent, who was responsible for creating the world. She is the giver and taker of life, and cannot be disturbed. Walking too close to the sandstone bluff is to risk upsetting the fragile truce between man and the spirits. If the serpent should ever rise, she would create a flood so large that the world would end.

The cliff is indeed a remnant from a time out of mind. There is no disagreement on this point.

Geologists say that approximately two hundred million years ago, the sandstone escarpment now known as Mount Brockman retreated from an original extension that had possibly touched the northern coast of Australia, some forty-three miles away. It withered away by degrees into its present shape, known as a Kombolgie Formation, with a layer of metamorphic sediment at its base. This mass was made up primarily of schist and muscovite and shot through with a matrix of cracks. A stream of acidic water started to flow beneath the fractured layer of sediments, carrying a liquor of dissolved heavy metals, and rose into the maze of Paleolithic fractures where the water evaporated and left behind patches of exposed uranium, which baked in the sunlight for millennia. The German explorer Ludwig Leichhardt, who passed near the site in 1847, reported: "I had a most disheartening, sickening view over a tremendously rocky country. A high land, composed of horizontal strata of sandstone, seemed to be literally 'hashed,' leaving the remaining blocks in fantastic figures of every shape and a green vegetation. . . ."

Mount Brockman remained unexplored by white men until 1969, when a two-seater aircraft passed just north of the mesa's highest point. The pilot and a technician were doing a radiometric survey for a company called Noranda Aluminum, Inc. When the plane banked away from the mesa, the spectrometer registered a massive spike in gamma radiation. Later, a team of geologists hacked its way in for a closer look and found what would then become the richest lode of uranium ever found in the Southern Hemisphere, an Australian version of Shinkolobwe.

Today, there is a chain of terraced pits at the foot of the holy Mount

Brockman: a mining colony that produces 8 percent of the world's uranium. I was taken around the Ranger Mine on a warm summer afternoon in January by a friendly public affairs official named Amanda Buckley.

"We're producing a commodity, just like coal," she told me. "Except this is just gray dirt. And at the end of the process we get dark green powder."

She showed me the giant pit where the pitchblende ore is blasted out, a fleet of haul trucks to move it out, a crushing plant to grind it into a fine sand, a series of leaching tanks to dissolve the sand into a watery state called pregnant liquor, a centrifuge to dry it, a giant oven to oxidize it, and a warehouse in which the finished yellowcake powder is sealed into drums painted red and loaded onto trucks going to the nearby city of Darwin for transfer onto oceangoing freighters and eventual use in nuclear reactors in Britain, South Korea, France, the United States, and Japan.

Ranger really did look like any other hard-rock mine and mill in the world, except for two things. Each employee was wearing a piece of filter paper encased in plastic on the lapel of his jumpsuit to measure the level of alpha-ray exposure. And there was a wooden stock fence on the southern perimeter of the mine. Any employee who crossed the fence without permission was subject to immediate termination. That was protected ground, the beginnings of the absolute no-go zone that encircles Mount Brockman.

Surrounding both the mine and the mountain is a domain called Kakadu National Park. Bigger than the state of Connecticut, it was chartered in 1977 as an awkward political compromise between conservative and liberal factions in Parliament. A holy mountain of the Aboriginals happened to be directly above (and was, itself, made of) a fortune in radioactive material. And it created a striking contradiction in land use: a working uranium mine in the middle of a national park.

This could have happened only in Australia, a place whose relationship with uranium has been nothing short of tortured. No other country has examined it as thoroughly: debates in Parliament, in countless newspaper op-eds, in church forums and in songs on the radio, and in the roadblocks and protests that closed down mine roads and office buildings and electrified national dialogue in the 1970s and again in the 1990s. The policies that limited the country's output have been relaxed in recent years, but Australia remains a lockbox of uranium. Nearly 40 percent of the world's known reserves are estimated to lie here, mostly unexploited.

Australia uses no uranium itself. The annual harvest of nearly nine thousand tons, once loaded onto ships, is never seen again. There are no atomic weapons, no enrichment facilities, and only one small nuclear reactor barely worth mentioning: a one-kilowatt facility in a suburb of Sydney used only for research. For all practical purposes, it is a nuclear-free country.

This has not mattered a bit. There was a time not so long ago when polls showed "uranium mining" was the top domestic policy issue, and almost no other topic had the power to start arguments and even fistfights throughout the country. Australians had serious questions about this particular mineral blessing, ranging from land use to fair taxation, foreign relations, environmental contamination, and possible nuclear war.

There was also the matter of race, which was a subject most people would have preferred to ignore. It remains a sensitive topic in Australia, almost as much as it is in America. But the uranium at Mount Brockman touched directly on the white majority's long and rocky history with the dark-skinned native people who had occupied the island continent for forty thousand years before the first shiploads of convicts arrived from Britain in the eighteenth century. The Aboriginals who wanted to save Mount Brockman tended to be viewed through the lens of a person's political beliefs. They were seen either as antediluvian whiners who needed to join the modern world or as noble martyrs to "progress," as embodied by all those uranium pyramids in the shadow of their holy site.

But nobody ever doubted their love and fear of the mountain. Archaeologists have found evidence of Aboriginal spirituality dating from six thousand years ago, which would make it one of the oldest religions on earth to be practiced continuously. Part of the supporting data is an etching of a piscine creature believed to be the Rainbow Serpent, which was recently found on a cliff not far from Mount Brockman. The image was likely painted there in 4000 B.C., near the end of the last Ice Age, when sea levels were rising from the melting North and South Poles, disrupting the ocean currents and washing unfamiliar fish onto the shores. These fish had apparently been viewed as signs from heaven.

One hundred fifty miles away from the sacred mountain, near the marina in the seaside city of Darwin, is a house with striped awnings. I went there on a summer afternoon in January to see a man named Joe Fisher.

Nobody answered the front doorbell, so I went around the back to knock on the carport door. There were two cars, some garden supplies, and a paperback Louis L'Amour Western novel tented up on a TV table near the door, as if whoever was reading it wanted to enjoy it in the warmth of the outside.

An elderly woman came to the door, smiling, and led me into a television room. She then went to rouse her husband from his postlunch-time nap. After about three minutes, Joe Fisher, eighty-eight years old, emerged from the bedroom with a bowlegged stagger and extended a hand.

He still sported the trim Walt Disney mustache that had been his trademark in the uranium fields half a century ago. He was an agent of every exhilarating thing that had characterized Australian uranium in its postwar years: huge profits, cowboy adventurism, and a fierce, even quasireligious, belief in progress. Everything that distinguished Charlie Steen in America also belonged to Joe Fisher in Australia. Today, he is regarded as an éminence grise of the mining business, and I was urged to go see him while he could still recall details.

"We were independent in those days," he told me. "No environmentalists to tell us what to do. We cut our own trees, built our own airstrips if we needed them. We would simply use a dragline behind a truck to tear the trees down. If you wanted anything done, you did it yourself. You didn't rely on anybody else."

He got into uranium not because he was fascinated with the mineral, but rather to make a little money. Fisher had spent his childhood in the 1920s chopping wood and welding engine parts for his father in the goldfields of Cape York, a dagger-shaped peninsula at the north end of the Australian continent. He learned early the basics of luck-and-sweat mining: shoveling tailings, lowering kibbles, and cutting down enough trees in one afternoon to produce half a cord of boiler fuel. After losing a finger to a generator fan, he was rushed three hundred miles to the nearest hospital and wound up marrying the nurse.

Fisher earned a degree in welding and drifted with his wife to one of Australia's ragged places: the Northern Territory, a province of jungle and arid scrublands called the Top End, a place with an end-of-the-road mystique similar to Alaska except that the climate here is tropical instead of freezing. The pace of life slows to a crawl during the summer, a period known as the Wet, when rains inundate the northern seaboard and turn roads to gumbo. The capital is a port town at the edge of the Timor Sea named Darwin, whose natural harbor was spied from the deck of the HMS *Beagle* in 1839 and named for Charles Darwin, who had been on the ship as a naturalist on a previous voyage. From here it is nearly one thousand miles down a bitumen highway to Alice Springs, the only other territorial town of consequence.

Darwin has a history of destruction and repurposing. On the morning of February 19, 1942, five days after Singapore fell to the Imperial Japanese navy, a squadron of nearly two hundred bombers and fighters unleashed a raid on the city's harbor and a nearby air base, sinking eight ships, destroying seventeen aircraft, and setting several storage tanks of oil ablaze. Most of downtown was flattened, and 243 people were killed. This was the first time Australia had ever been attacked by a foreign enemy. The ash-covered remnants were hit by Japanese air attacks multiple times before the end of World War II.

Darwin's economic base was wrecked, and the federal government was reluctant to pay for the reconstruction of such a remote outpost. Downtown buildings were left as shells well into the next decade. In 1974, a giant Pacific typhoon named Cyclone Tracy made landfall near the harbor and leveled the city once again. In the years between these two obliterations, Darwin struggled to create more jobs and opportunities beyond the docks and the rail yards. Few people wanted to move there. Beer, vegetables, and clothes had to be trucked or flown in; most of the milk was powdered.

In place of a sewer system, the residents used pit toilets known as flaming furies—metal drums half sunk into the ground and in which human waste was burned out with diesel fuel in the evenings. The nightly stench of the burning toilets was known locally as *la perfume*. There were no operating funeral homes, and the territorial morgue had no electricity. A new employee at the morgue created a commotion one night in 1954 when he saw sweat beads forming on a corpse's skin; he thought the body had come to life. But the "sweat" turned out to be condensation from the muggy air.

One of Darwin's residents was a prospector named John Michael "Jack" White, who in 1949 started rummaging around some old copper shafts near a spot called Rum Jungle. The place had been named in honor of a nineteenth-century wagon crew that got bogged down in mud on its way to deliver some rum kegs to a cable station. Nervous about crocodiles in the streams, the teamsters decided to drink the rum. There had been several hundred gallons at their disposal, and the party lasted for days. Near this spot, Jack White found greenish rocks that clearly were not copper, but they gave no other hint as to their identity. Geologists in Australia had been furnished with guidebooks picturing the various expressions of uranium—carnotite, pitchblende, brannanite, coffinite, and the like—and White recognized an oxide called torbernite. Contractors redug the copper tunnels to five hundred feet and sealed off the area. The uranium went to Britain's atomic arsenal. Jack White was rewarded with a $50,000 finder's fee, though there were complaints in Darwin about the low amount he had been paid for such a find.

The news still created a jolt. Here at last was a source of ready cash in the Top End. It did not seem to matter that few—if any—Australian geologists knew a thing about uranium. The federal Bureau of Mineral Resources issued topographical maps, printed how-to manuals, intervened in claim disputes, and assayed samples. This was almost identical to what was happening in America at the same time: Ordinary people were encouraged to scour a desert in exchange for the possibility of a large cash reward—an odd melding of capitalistic incentive and state oversight; Wismut and Utah together again. It was what economists call a monopsony: a single buyer and a lot of sellers.

Joe Fisher showed up in the midst of this gambler's atmosphere attached to the company United Uranium No Liability, a job he acquired through family connections. This gave him a paycheck and credibility in a time of shifting fortunes. UUNL had a reputation as a well-financed outfit, with top geologists on the payroll, and it would later go on to develop fourteen productive mines in the South Alligator Valley. But nothing was certain then. One of Fisher's first jobs was at a bluff-side claim called El Sharana, where a huge lode of uranium had just been named for the three young daughters—Ellen, Sharon, and Anna—of the chief prospector.

Joe Fisher found a headache, however. The road to the claim turned out to be so precipitous that heavy-torque bulldozers had to winch the trucks up the bluff. The camp itself was also a maze of canvas tents connected to the highway via a rough road through a crocodile-infested river. Fisher fired the caretaker, ordered a new camp constructed, and started bulldozing a new road. Mosquitoes made black funnel clouds in the evenings, and the average daytime temperature in the summers topped 100 degrees Fahrenheit. The camp had a movie projector, but only two films: a drama called *Westward the Women* and a documentary, *Erotic Art of India*. The miners watched them over and over. There was nothing else to do. "They were long, tough days," Fisher recalled, "but there could be no turning back."

Another problem was the airstrip, which had been foolishly oriented perpendicular to the mountainside. When the wind was wrong, the planes were forced to take off aimed straight for the cliff and then execute a hairpin turn as soon as they were aloft. This was judged too risky, even for the uranium daredevils. Fisher had just started ripping out trees for a new airstrip when an official of the Department of Civil Aviation told him the entire site was unsafe and refused to grant a license. Fisher would have no choice but to find a landing site farther away. The access road would have to be more than thirty-seven miles long and cross the twisting South Alligator River twice. The road would also pass by Coronation Hill, an equally promising strike nearby that had been discovered on the same day that Queen Elizabeth II was crowned in London.

Heavy rains pounded the area in the Wet of 1955, washing away the fill from the causeways and turning the road to muck. Fisher had to borrow two army trucks and a winch to haul uranium over the roads; the river crossings were served by a government truck with axles that could clear the water, as well as a "flying fox"—a suspended cage attached to the trees with a rope, in which men and equipment could be pulleyed across the river.

He managed to open the road and airstrip by the beginning of the next "Dry." There was also a bonus. While flying the route one day, he did a spectrometer survey (there was always time for that) and found a radioactive patch on another mountain slope, rich in pitchblende, that produced a rainbow smear of colors on the incline. Fisher named the site Palette after the artist's tool and ordered a corkscrewed road built up to the site to start exploiting it.

Fisher had been among the lucky ones. He had access to a real company with real money. For the less-well-connected locals in Darwin, there were only two ways to get in. You could go out into the steaming jungle and wave around a Geiger counter. Or you could buy penny shares in one of the hundreds of companies forming inside living rooms and over glasses of Tooheys Draught. The bar at the Darwin Hotel had become an informal stock exchange.

A man named Ross Annabell decided to go prospecting after he got fired from his job at the *Northern Territory News*. Investing didn't excite him. "New offices were opening, new base camps being set up," he recalled in his memoirs. "Geologists fresh from the south were racing around in Land Rovers, and prospectors were being recruited. There was a smell of money in the air."

He soon became disheartened. Roads out of Darwin were often impassable or nonexistent. As it was in the United States, the hunting was all done on the vast reserves of unsettled government land, much of it rough and isolated. Swarms of insects crawled in a prospector's eyes and buzzed in his ears at night. And there was the feathery red soil called bull dust, which could bog down a jeep just as badly as mud had trapped the rum-drinking teamsters of the previous century. Complained Annabell: "It gets into your hair and lungs, your clothes, your camera, into the innermost recesses of your gear (no matter how well you roll your swag), and into your eyes and into your food." Those who lost their tempers in the bush were said to have "gone troppo," or been driven crazy by the stultifying tropical air. A few took to blasting away with shotguns at logs floating in the creeks, believing them to be crocodiles, which sometimes they were. Alcohol, of course, was a dependable refuge. The road east of Darwin became known as the Glass Highway for the countless empty bottles of beer and whiskey tossed out of car windows.

Annabell's tools were a pick and a shovel and a Geiger counter with a pair of headphones. He partnered with an Aboriginal named Dick, who occasionally worried out loud about whether he was inadvertently treading on sacred ground and thus potentially triggering the catastrophe that could end the world. Annabell described Dick as a "mixture of stone age and atomic age," which, in a way, summed up the whole enterprise.

They employed a crude method for inspecting ground. Each morning, they would ride into the bush on horseback and toss lighted matches into the grass. The men would then double back after the resulting wildfires had burned away the top covering and pass their Geigers over the ash. Anything that moved the needle was pegged, which was a task in itself. Inexperienced prospectors could be easily fooled by streaks of potassium or thorium in the slate, or even the radium on their own watch dials.

It was a chimera for most, this jungle uranium, but all the false readings and dry holes did nothing to discourage the frenzy. The uranium had a peculiarly *female* quality to it—an allusion made time and again by the Australian wanderers who found themselves bewitched by quivering needles. One lucky finder named his mine Fleur de Lys, after an attractive woman named Lys, who worked as a cocktail waitress at the Darwin Hotel. One of Australia's largest uranium mines would be named Mary Kathleen. Said Ross Annabell: "Uranium is an unpredictable lady, every bit as fickle as gold. That she flowers on the surface with the rich greens and yellows of secondary ores does not necessarily mean she's there in depth. Success lies in the location of the primary ore body, which may not exist." Those minerals may have eroded away, he said, leaving nothing but a gorgeous smear of carnotite as a tease.

For Joe Fisher, these were the headiest days of his life; he was on the trail of an exotic mineral, as his father had been; uranium was the material that would shake the Top End out of its slumber and raise it to the top echelon of Australian states. He had found the adventure of a lifetime with UUNL. His mantra, like that of many other of his contemporaries, was simple: "Wealth follows energy."

Most of the fly-by-nights were out of business by 1960, but the established producers were reaping a rich harvest, exporting $164 million of yellowcake each year. The El Sharana mine yielded the biggest piece of solid pitchblende ever recorded—nearly one ton. The black blob was hauled to the Darwin fairgrounds and put on exhibit as a publicity stunt; newspapers across the country ran photographs. UUNL had become wealthy enough to acquire the neighboring Coronation Hill site and built a large gravity mill at El Sharana. Fisher's wife, Eleanor, was the one chosen to push the ceremonial button setting the mill to life.

The couple was invited to dine at a banquet with Queen Elizabeth II herself when she docked the royal yacht *Britannia* in Darwin. After drinking three martinis to calm his nerves, Fisher took his assigned seat next to the sovereign. Over a meal of lobster and spaghetti, Queen Elizabeth turned to him and talked for thirty minutes about uranium hunting. In a note to himself scribbled immediately afterward, Fisher wrote, "Her voice is beautiful, her hands expressive. In conversation, her eyes light up and one feels a real interest is being taken in the discussion."

He had a deeper reason to feel that way. The defense policies of Her Majesty's government were directly responsible for his fortune. The British Atomic Energy Authority had been contracting with his company for the production of two thousand tons of uranium oxide a year. A portion of it was enriched to a level above 90 percent U-235 and placed in the core of atomic bombs.

That uranium would be returned to its home country, after a fashion, through a series of nuclear tests conducted on a southern plain called Maralinga. Seven nuclear weapons were exploded there by the British military in the 1950s, leaving radioactivity in the soil and in the lungs of Australian soldiers assigned to monitor the site. These events would come to stir popular resentment toward Great Britain, whose treatment of its antipodal cousins had been a source of tension ever since the battle of Gallipoli in 1915, when Australian troops were ordered into a hopeless attempt to capture a Turkish peninsula. Many Australians felt their newest mineral export was becoming a means for their nation to be treated, once again, as a cash box and garbage bin by the colonial power.

Joe Fisher remained undaunted. Uranium was like gold to him: the mineral that could build up Australia and make it rich among nations.

"Why should we have left it in the ground?" he said. "It is there to be used."

The hazardous qualities of the mineral—either the immediate danger of alpha rays or the more abstract worry about its use in nuclear weapons—never bothered him.

"I'd sleep on a bed of uranium and wouldn't worry a bit about it," he told me. "You don't need it to make bombs. Ordinary gunpowder would do the job just as well."

But the good times were drying up. The five atomic superpowers—Great Britain, the USSR, France, China, and the United States—slowed down their crash acquisitions just as the uranium machine had been revved up to a frenzy. Yellowcake barrels began to stack up in the warehouses; companies were forced to sell their inventories at a deep discount.

A key buyer was absent. The nuclear power industry was taking a painfully long time to come online, despite the blue-sky predictions of William L. Laurence and futurists who followed his lead. Total worldwide capacity was just one gigawatt by 1960, barely enough to light one small city for eight hours a day. The United States announced it would shield its own companies from bankruptcy by enriching only the uranium that came from its own Western deserts, closing off the choicest market for the suddenly abundant mineral. Production had dropped by 40 percent by 1965, but the price kept plummeting.

This was a period of market evolution when only one thing was known to everyone: A Thing this destructive could be traded like no ordinary commodity. Only military superpowers and a few public utilities could buy it. And they were free to deal with a mélange of suppliers, ranging from energy giants such as Gulf Oil, Phillips Petroleum, and Rio Tinto Zinc to government marketing boards such as the French cooperative Uranex, the South African Nufcor, and the Canadian UCAN, which had acquired their stock from more low-level players such as Joe Fisher's UUNL and Charlie Steen's Utex. The contracts were long term, typically ten years. Terms and prices were dictated by national security decisions, made in secret. All of this made the uranium market about as flexible as a tree stump.

The small circle of buyers and sellers dealt with one another directly and, at times, through a consultancy in Menlo Park, California, called the Nuclear Exchange Corp., or Nuexco. It had been formed to be a bank for uranium, but evolved into a brokerage that published a list of recent prices. By 1971, that price was on the verge of dropping below $5 per pound, and the mining companies sensed a disaster in the making.

They decided to have a secret conference of their own. Mining executives from UCAN, Nufcor, and Rio Tinto, as well as government representatives from Canada and Australia, gathered inside the Paris offices of the agency Commissariat à l'Énergie Atomique on February 1, 1972, to have an unusual conversation. Buyers were not invited nor was the U.S. Atomic Energy Commission. These discussions were illegal under American antitrust law. They were also for the purpose of screwing the buyers, and everybody there knew it.

After coffee and pleasantries, there were admonitions to mind the "extreme secrecy of the business discussed." Then a system of collusion was proposed. When a producer learned of a request to buy uranium, it would inform the French agency, which would pick a supplier to offer at 8 cents under. The deals would be rotated evenly among members. Each would be guaranteed contracts at a fixed price, eliminating the need for harmful competition.

"The system was based on the French metaphor of the filling of wine-glasses," said one observer. "Each country had a glass to be filled; when it was full, it was somebody else's turn. If one glass was not completely filled, it would have access to the next round until it was."

The point man at the French agency was the discreet secretary André Petit, who agreed to serve as head of the "research organization," which soon took on a more flippant, if more accurate, nickname: the "Uranium Club."

In the initial meetings, the club set a floor price of $5.40 a pound, calculated to keep every member solvent. The Australians, in particular, said (accurately) they were sitting on much bigger reserves than the others and deserved an equitable slice of the pie—especially after 1977, when demand was expected to pick up from new nuclear power plants. "Australia hoped to play the role in the Club that Saudi Arabia plays in OPEC," noted one analyst later.

The divvying was not to everyone's liking, at first. At a conference in Johannesburg later that year, it was finally agreed that the Canadians could have 33.5 percent; the South Africans, 23.75 percent; the French, 21.75 percent; the Australians, a flat 17 percent; and Rio Tinto Zinc, 4 percent.

The presence of the Rio Tinto company among this breadline of sovereign nations was a reminder of just how incestuous the uranium trade had become. It also demonstrated the matchless reach of Rio Tinto, which tended to behave as though it was a wholly owned subsidiary of the British throne. Many believed it was exactly that. One of the major shareholders was supposed to have been Queen Elizabeth II herself, via a secret account at the Bank of England. The *Times* of London once remarked, apparently without irony, that it was "almost patriotic" for an ordinary Briton to own shares in the company.

Rio Tinto had been founded in 1873 as a venture backed by the Rothschild family to restart some abandoned shafts in southern Spain that had once supplied the Roman Empire with copper. But it did not become a global superpower until the 1950s, when it came under the chairmanship of the urbane Sir Val Duncan, a Royal Engineer during the war and a director of the Bank of England. He built up a network of railways, ports, and mills to extract minerals from Britain's former colonial possessions. Its web of affiliate companies was a closely guarded secret, its ownership records kept inside a four-inch-thick book known within the company as the Bible.

Most important, Sir Val hired a series of executives and recruited board members with close ties to Parliament and the Foreign Ministry.* Sir Val's sense of Rio Tinto (and perhaps himself) as a shadow version of Parliament is best illustrated by a remark he made at a dinner party in

---

* Star employees were former prime minister Anthony Eden and Lord Peter Carrington, who would go on to serve as defense secretary and secretary of state for energy.

1974, when confidence in Prime Minister Harold Wilson was at an all-time low. Some feared Wilson might actually be overthrown in a coup, but the possibility did not worry the lion of Rio.

"When anarchy comes, we are going to provide a lot of essential generators to keep electricity going," said Sir Val Duncan. "Then the army will play its proper role."

Rio Tinto had been quick to join the nuclear revolution and now felt itself overextended. Its uranium holdings were spread throughout South Africa, Namibia, Canada, and even Australia, where it had a stake in the Mary Kathleen mine. Some investigators later suspected that Rio Tinto itself, acting through a Canadian subsidiary, was the initiator of the secret talks in Paris.

Dividing up the world's uranium through a gentleman's agreement would not have been a particularly novel idea for Rio Tinto. Its leadership was intertwined not only with the ruling class in Britain—where atomic stewardship was now a four-decade-old tradition—but also the boards of many of its competitors. Sir Val, for example, had a seat with a French company called Imetal, which had a controlling stake in a Gabonese uranium producer. On that same board was Harry Oppenheimer, the chairman of Anglo-American, the largest producer of uranium in South Africa and a legendary mining omnivore. (Oppenheimer was also the chairman of the diamond wholesaler De Beers, an entity that happened to know a thing or two about price-rigging.) As it happened, an 8 percent stake in Rio Tinto was held by Charter Consolidated, which was in turn owned at 36 percent by Anglo-American. Several charter members of the Uranium Club had thereby been doing business with one another long before they met in Paris.

If the oligarchic nature of the club bothered any of them, they did not show it. Nor did they ever break their promise and discuss their activities in public. Power companies suspected what was really happening behind the curtains, but were forced to keep buying $8 uranium if they wanted to keep their turbines running. "I infer they are setting prices because their asking prices always seem to come out the same, if you know what I mean," an anonymous employee of Nuexco told *Forbes*.

A Canadian member of the club was more direct. "It worked for the

Arabs, didn't it?" he said, referring admiringly to the Organization of Petroleum Exporting Countries (OPEC) oil cartel, then in its formative days.

But in its own way, this rent seeking was not substantially different from Edgar Sengier's Union Minière, which had depended on the favor of the Belgian and U.S. governments and a ring of elite mining interests to preserve its early lock on the world's supply of uranium. It was merely a corporate version of what Leslie Groves had anticipated when he first told his geologists to assess the likelihood of a future American hegemony.

The Uranium Club operated in this fashion for three years until a combination of factors contributed to its demise. The United States announced it would soon end its trade ban on overseas uranium. At roughly the same time, OPEC began an oil boycott, sending energy markets into a panic. Nuclear power was looking like an attractive alternative to oil, and the price of uranium began to take off. By 1974, the price for a delivery nine years in the future had climbed to $23 a pound. Geologists were sent back into the field. The need to collude had vanished, and producers began once again to bid one another down. The club disbanded the next year, later reconstituting itself in London as a research and advocacy group with the more benign name Uranium Institute.

The Paris meetings would have faded into legend were it not for the actions of a whistle-blower. A disgruntled employee—never identified— of the Mary Kathleen mine dropped off a box of incriminating documents one night on the doorstep of a lobbying group called Friends of the Earth, whose members spent a night photocopying the materials before turning them in to the police as stolen property the next morning.

These papers later became exhibits in a chain of lawsuits filed by the plant-building giant Westinghouse, which had guaranteed its customers twenty years of uranium at bargain-basement prices and then stood to lose $2.5 billion when the price started to climb. This blunder has been called one of the worst mistakes ever made by a major American corporation. Westinghouse grew so desperate to fulfill its contracts that it began experimenting with "purification plants" to wring trace amounts from slag heaps all over the American West—whether or not the neighboring mines had anything to do with uranium.

"Oh, they were in a terrible pickle," said Doug Duncan, who was one of the managers of the purification plant at a copper mine outside Salt Lake City. "It was desperation, you might say. They were up to their ears in obligations to provide uranium and were trying to squeeze as much of it out of the ground as possible."

The denouncement took place in courtrooms. Westinghouse officially blamed the club and unsuccessfully tried to subpoena top executives of Rio Tinto and other cartel members, who objected on convenient grounds of "nuclear security." The cases eventually settled, and the purification plants closed down.

But there was another reason the club fell apart. Back in Australia, the attitude toward uranium was about to take a turn. Campaigning with the slogan "It's time," the left-leaning Labour Party won a majority of seats in Parliament and began to reverse two decades of Conservative policy. Prime Minister Gough Whitlam began rethinking uranium. The government suspended all exports and forbade the opening of new mines, even as prices were spiking to new highs.

Australia then became embroiled in a question that had never been asked so loudly or so insistently since the first confused days after Hiroshima: Is uranium immoral?

The flashpoint had been Mount Brockman. A company named Energy Resources of Australia had acquired the lease to the Ranger site and was preparing to dig when the government suddenly announced the creation of a national park named Kakadu. It would encircle 7,700 square miles and incorporate a large part of the South Alligator River uranium fields, including the Ranger Mine. This would have shut down virtually all production in the Top End and turned Australia into a nonplayer in the world uranium business.

The Labour Party didn't campaign on this issue, but its parliamentarians had been receiving multiple visits and phone calls from organizations such as the Friends of the Earth Australia, whose members were a colorful crew of college professors, anti-Vietnam protesters, and veterans of the ragtag Greenpeace campaigns that had challenged French nuclear testing in the Pacific. A key part of their strategy was handing out agitprop pamphlets that summarized the objections to uranium—its radioactive waste, its disregard of Aboriginal land, its contributions to nuclear proliferation and the possibility of global destruction. India's first atomic bomb had recently been test-exploded with the help of Canadian uranium. Was putting all this stuff into the world really making the planet any safer?

Friends of the Earth called on Australians to forswear the money and leave the uranium in the ground. "We have a duty to care for the Earth, our home," said one of their handouts. "We are responsible for our exports and their ultimate impact on the environment." The government agreed to form a judicial panel to decide the fate of Kakadu National Park—and with it, the future of the northern uranium business.

Critics were livid. Even if Australia were to withhold its treasure, the stuff could easily be mined elsewhere, and overseas companies would be delighted to fill the gap. There were existing mineral leases, such as Ranger, to be considered, as well as the loss of billions of dollars in revenue and thousands of steady jobs for working-class Australians. Some suspected that uranium was merely a cover for larger socioeconomic grievances. Mining executives, in particular, disliked being called warmongers and waste barons.

"The debate in Australia was assuming some of the characteristics of a class war," wrote Tony Grey, founder and chairman of the uranium company Pancontinental. "Uranium mining was being clothed in materialism and identified with the gaudy rich. Money loomed in conflict with morality."

But the traders had done themselves no favors by creating an alarming image around their product. Trucks taking yellowcake barrels to port were often accompanied by police vehicles, their lights whirling, closing down major intersections and making a spectacle of what ought to have been a banal transfer of goods. Railroad workers were persuaded to walk off the job for one day, in protest of uranium. "I wish the bloody stuff had never been discovered," said the union leader Bob Hawke, in an unguarded moment.

The price of uranium, meanwhile, had gone up nearly seven times from what it had been in the heyday of the club. Australian mining companies were powerless to increase their production; millions of dollars were lost. "We felt time slipping away," recalled Gray. "The uranium market was booming and we could do nothing about it."

Joe Fisher was especially angry. "Never in this modern age has there been a precedent where a national park of exceptional magnitude was declared over mineral rich geological structures containing one of the largest potential energy and mineral resources known to mankind," he complained in his memoirs.

He also ridiculed the antinuclear demonstrators. "They were handing out literature, stating that miners should desist from blasting rocks as they were capable of feeling pain," he wrote. "They are unreal. Even if they were silly enough to believe that rocks feel pain, why are they not picketing the quarries that abound around large cities in the States?"

There was a more personal element for Fisher. He felt, as did Tony Gray, that uranium miners were being cast as land-raping villains. This insulted all the pluck and bravery of the first uranium pioneers. If it wasn't for the drive to build atomic bombs, he said, the Top End would still be deserted. The lowlands of Kakadu were already crisscrossed with mine roads and pockmarked with diggings, many of which had been put there by Fisher himself.

"Mineral explorers alone had the imagination and appreciation to take the first steps in bringing the area's natural grandeur to public attention," he wrote.

A more reasonable compromise, he argued, would be to set off an area of no more than twelve hundred square miles—not a mammoth blanket of public protection thrown over the entire mineralized area with so much uranium there left to use and sell. He called Kakadu "the park that grew too fat" and was gratified when the famous British naturalist Kenneth Mellanby deemed it "mongrel country" and "scruffy" in an interview with a Darwin newspaper.

"I enjoyed my visit, it was very interesting, but to the ordinary person it must be the most boring national park in the world," said Mellanby.

After interviewing 287 witnesses and compiling more than twelve thousand pages of evidence, the Ranger Uranium Environmental Inquiry came out with a report that was tied up in ideological knots. It approved the creation of Kakadu National Park, but upheld the Ranger lease *at the same time*. This meant there would be a West Berlin–style mining colony as an industrialized island in the middle of the park, surrounded on all sides by wilderness. A company village could also be built a respectful distance away from Mount Brockman. Local Aboriginals would have the right to veto any future improvements under the new Land Rights Act of 1976.

The report came laced with a whiff of self-pity. "There can be no compromise with the position: either it is trusted as conclusive, or it is set aside. We are a tribunal of white men and any attempt on our part to state what is a reasonable accommodation of the various claims and interests can be regarded as white man's arrogance or paternalism. Nevertheless, this is the task we have been set."

Here was the heart of it, the national wound of Australia—the historically wretched treatment of the dark-skinned people colloquially known as blackfellows, but more properly called Aboriginals, whose Dreamtime places were soon to be spaded up for radioactivity.

They had lived here as nomads for more than forty thousand years, growing no crops and taking their sustenance from scavenged berries and insects, as well as game and waterfowl that they hunted with spears. Initiation rites were complex and lengthy. Land was held in common by members of an extended clan, in conjunction with a story detailing the holiness of a particular cliff or river, often the dwelling of a spirit creature. The story was like having a lease on the property. It often came bundled with apocalyptic visions, as the penalty for treading on a restricted spot was usually personal disaster or the end of the world.

They feared the country, but loved it, too. The words for "land" and "home" are the same in most Aboriginal dialects. A song from the Oenpelli region:

> *Come with me to the point and we'll look at the country*
> *We'll look at the rocks*
> *Look, rain is coming*
> *It falls on my sweetheart*

The arrival of British convict ships in the 1780s had been a catastrophe for the Aboriginals. They had no immunity to smallpox, measles, and flu; huge numbers of them died within weeks of their first contact with the whites. Those roaming in the deserts of the interior, and the tropical savannas near Mount Brockman, were insulated from the first wave of infections, but as the interior began to fill up with British cattlemen in the 1800s, life grew even more difficult. The white settlers tended to view the Aboriginals as either low-wage laborers or outright impediments to opening up the new country; their passionate connection to the land was ignored. They were called idle, lazy, and vengeful. "They are given to extreme gluttony and if possible will sleep both day and night," complained one settler in a letter home.

Occasional massacres erupted; as many as twenty thousand Aboriginals may have died violently at the hands of whites over the years. They were sometimes given liquor as a joke and encouraged to knife each other. Reported a Protestant minister named Reverend Yate: "I have heard again and again people say that they were nothing better than dogs, and that it was no more harm to shoot them than it would be to shoot a dog."

The practice of taking Aboriginal girls as concubines—often against their will—was common at many outback cattle stations. This had even more tragic consequences. Believing the indigenous race was in decline, the federal government instituted a "child removal" policy in 1915 in which the offspring of mixed-blood unions were taken from their parents and housed in orphanages, internment camps, and foster homes with the aim of assimilating them into white society. They were often taught to believe that their real parents were stupid and shiftless. Few Aboriginals were allowed to become Australian citizens; those who were so privileged were required to carry identification papers, known colloquially as dog licenses, to prove it.

By the middle of the twentieth century, the nomadic life had all but vanished. Aboriginal numbers were less than a tenth of the half million or so who were alive before the British settlers landed. Most were living on the fringes of society, suffering the same alienated fates as the American Indians of their day—alcoholism, disease, and nihilism. The uranium bonanza had pushed them further to the margins, especially in the Top End, where many lived on cattle stations or at missionary-run charities. Some took to vagrancy, begging for change. They were finally granted the right to vote in 1963, but they were not allowed in many swimming pools or hotels until much later. The compromise over Ranger was viewed as a kind of payback for all the abuse. The Aboriginals now—suddenly—had an exceptionally strong hand in deciding how uranium was to be developed, if at all.

This puzzled Joe Fisher, who had employed dozens of Aboriginals in mines and mills, and felt they should not have received any superior status in regards to the land. "One cannot dispute that their lifestyle is different from that of the white man, that their customs are different and their culture is different," he wrote. "Yet, essentially they remain human beings just like you and me; and they should be treated just like you and me as human beings, not differently."

He was quick to profess respect for their ways. During one jeep trip near Port Keats, he had watched as an Aboriginal companion caught a goanna lizard that he wanted to take back to his family for dinner. He could not kill it on the spot—its corpse would grow rotten in the heat—so he snapped the four legs and threw it alive in the back of the truck. The reptile "remained fresh until the feast," noted Fisher, admiringly.

He added: "Provided you worked with them, they were generally good hands."

The company agreed to pay the local clan of Aboriginals, known as the Mirrar, a royalty on gross sales, plus an annual rental fee. It also agreed to move its southern boundary even farther away from the sacred Mount Brockman. An artificial town named Jabiru was platted and constructed in the middle of the tropical forest. It had a school, a supermarket, a gas station, a cul-de-sac neighborhood of ranch houses, a Holiday Inn in the shape of a crocodile, a camping spot for wandering Aboriginals, a mimeographed local newspaper called the *Rag*, and wide bitumen streets that shimmered in the afternoon heat and seemed to fade into the trees. Democracy was another mirage: a Ranger executive was a permanent member of the city council. Unless you worked at the uranium mine, you could not stay there. On the streets at shift change, miners in orange jumpsuits, helmets, and work boots could be seen walking from the shuttle bus stop to their homes. The nearest shopping for anything but groceries and postage stamps was in Darwin, 140 miles away. Jobless Aboriginals lingered outside the concrete-block supermarket, occasionally cadging for pocket change. The highway to the mine had a panoramic view of Mount Brockman, where nobody was allowed to tread.

The town was a mirror of the tortured national attitude toward uranium. Few people wanted to stay there when their time at the mine was finished. The Kakadu park itself was rarely visited by tourists and was viewed as an unattractive compromise. An environmental group called it "a controlled disaster zone rather than a national park." The historian David Lawrence settled on this verdict: "a mélange of uranium mining, environmental conservation, tourism, and Aboriginal land rights."

For atomic proponents, it was regarded as an acceptable—if ungainly—outcome. Though he hated the vast size of the park, Joe Fisher defended the decision to carve out the Ranger lode.

"Some conservationists have claimed that to allow any mining whatsoever in an attractive place like Kakadu is like cutting a small hole in a Van Gogh painting," he said. "This is a totally misleading analogy. One looks at a painting as a whole; no human can take in an area as large as Kakadu." In the Dreamtime country, the uranium existed only for those who wanted to see it.

Mount Brockman was not the only patch of uranium in the strange new park. An even richer deposit called Jabiluka had been discovered in 1970 by a geological team sent on a joint venture of Pancontinental and Getty Oil. When the drill samples showed the presence of uranium oxide, the chief geologist sent a one-word telegram to his bosses: CHAMPAGNE! As mining permits were being finalized in 1983, the political climate shifted once more, and Labour was voted back into power with a much less murky philosophy toward uranium. A grandfather clause, otherwise known as the Three Mines policy, froze the business in place. The operations then in existence—Ranger, Mary Kathleen, and Olympic Dam—could continue. But no further export licenses would be approved. This looked like Jabiluka's obituary.

But in 1991, the lease was sold to ERA, the next-door operators of

Ranger, who restarted the approval process for Jabiluka and had better luck with the politicians in Canberra. Further test drilling indicated that the lode was much bigger than previously thought. Once milled, it would yield nearly 120,000 tons of yellowcake. However, protesting the "rape of Jabiluka" quickly became a fashionable cause among young Australian progressives, even among those who had never visited the Top End and had no intention of ever going. In Sydney, five thousand people gathered to picket in the streets; three thousand were in Melbourne the same day. A Catholic advocacy group labeled the uranium project "morally unacceptable."

At the rallies rock bands performed songs about, in the words of one musician, "the death and destruction that is inherent to uranium mining." Peter Garrett, the lead singer of the rock band Midnight Oil, was the headline act at several rallies. He later used his role in the protests as a springboard for a successful campaign for a seat in the national Senate. Outside the locked gates of the Jabiluka site, he told a teeming crowd: "Any fair-minded Australian who had thought the issue of having twenty million tons of radioactive tailings in a World Heritage listed site, in the middle of the most significant national park we have, on land that belongs to somebody else, will say that this mine is wrong."

A Midnight Oil song called "The Dead Heart" later received worldwide radio airplay and denounced uranium companies. The chorus appropriated the supposed voice of the Aboriginal:

> We carry in our hearts the true country
> And that cannot be stolen
> We follow in the steps of our ancestry
> And that cannot be broken

A tent city sprang up at the gates. Local police requested a $1 million increase in their budget just to keep order. Sleeping on the roadside there became a mark of prestige. Animal blood was splashed on bulldozers, and holes were cut in the fence at night. A van crashed through the entrance to the mine site; the driver was arrested. In apparent retaliation, the driver of a twenty-four-wheeled truck from the mine came roaring through the protesters' campsite, knocking over several tents. He claimed to have "gotten lost." Molotov cocktails were hurled through the windows of ERA's offices in Melbourne, causing fire damage but no injuries.

For many Australians, the uranium hubbub cut to a core question, one of national identity: What does it mean to be an Australian? Do we stand for globalization or isolation? What responsibility do we have to the rest of the world?

In December 1998, two protesters in Melbourne were arrested for spray painting antiuranium slogans on a statue of Robert O'Hara Burke and William Wills, heroes of the previous century who scoured Aboriginal land for gold. These men were a Down Under version of Lewis and Clark, opening up territory for progress and civilization. Uranium was squarely within that heritage of Australian discovery. But it was a different kind of treasure from gold—more mysterious and sinister, coming to the forefront of the national agenda in a more prosperous era in which the majority of citizens lived in air-conditioned houses, drove cars, ate well, watched television, and therefore had the luxury of refusing to develop a natural resource that represented fabulous profits. This wasn't Niger or the Congo or even East Germany in the 1950s, where uranium was all that separated rural people from abject poverty. Australia could say no because it could afford that luxury. "Most Australians don't want it, the traditional owners don't want it, the world doesn't need it," said David Sweeney of the Australian Conservation Foundation. "It's unsafe, it's unclean, it's unnecessary."

The protesters received a boost in November 1998 when the United Nations sent a seven-member team of experts to the area and reported that the mining would pose an imminent threat because of the radioactive tailings. Two years later, a majority stake in ERA was purchased by the British mining giant Rio Tinto (that stalwart of the old Uranium Club), which announced the site would not be developed, given the fierce political opposition.

I visited the second-story offices of the Environment Center of the Northern Territory on a morning in January when the air was warm and smelling of wet steel, a sign of an imminent monsoon. The air conditioner was broken, and everyone inside was sweating. A framed newspaper on the wall, from a decade earlier, carried the headline GREENIES WARNED: BEHAVE OR ELSE.

A staff member named Emma King made us hot coffee in a press jar. She told me she feared the energy had gone out of the antinuclear movement and that Australia was fated to become a bigger uranium producer because the opposition had become lazy.

"People are less willing to engage anymore," she said. "It's too overwhelming for them."

King is a suntanned woman of forty with a tattoo on her forearm in the shape of a dog's paw print, a memorial to her lost blue heeler, Molly, who had run away the previous year. King had moved up to Darwin to be a reporter and wound up as the head uranium campaigner for the Environment Center. She had come to believe it was immoral to dig radioactive material out of the ground, even for a purpose as benign as generating electric power. After citing a list of objections for me—hazardous waste, scars on the land, theft of Aboriginal land—she brought up a more personal critique.

"It's really hard to put this idea forth without being viewed as a conspiracy theorist or a Marxist," she said. "But I really don't understand why our money isn't being spent on developing renewable sources of energy like wind or solar power. We will eventually have to go to renewables. Why don't we do it now? My idea is that this would go against interests of multinational companies because they cannot control the source of energy. Nobody owns the wind or the sky. But with nuclear energy, you can control the uranium. That means you can set whatever price you want. Nuclear seems to be the easy answer to climate change, right? It's going to solve all our problems and people won't have to take personal responsibility. But I think it's a way for corporations and governments to retain control over our energy supply."

What about the uranium itself, I wanted to know. Could it ever be used for a good purpose?

"This isn't necessarily my theory, but I've heard people speculate that maybe it's a part of the evolutionary process," she said.

I asked her to elaborate.

"Radioactivity leads to mutations. Evolution needs mutations." Man's tinkering with uranium, she went on, could lead to unexpected genetic changes in the species itself. The mineral had lain undisturbed and unnoticed in the Congo, in Utah, in Australia, and in eastern Germany until civilization reached a certain point in the first half of the twentieth century. Uranium was then "discovered" by the physicists, and all of its latent power became known. Hiroshima happened and rearranged the globe. And through either a catastrophic atomic war or just the incremental effect of mutant-making waste piling upward, uranium would give birth to a new version of man—just as surely as a scarcity of food on the Galápagos Islands had forced Charles Darwin's mockingbirds to adapt and evolve.

In this theory, uranium plays a role like the black monolith in the science fiction film *2001: A Space Odyssey*, which lay buried on the moon like a time capsule until man was knowledgeable enough to travel to the Tycho Crater and detect the enormous radio waves coming from the object. When the monolith was exposed to the light of the sun for the first time, a new era of evolution and a new chapter of mankind could begin.

But uranium might also be seen as a serpent out of John Milton or a rough beast out of Yeats, a sentinel of dystopia, the apple of knowledge force-fed to the unready, who are exiled into a world they never asked for and do not want. A Jabiluka protester told me he had joked with his friends while looking at the night sky: "Each one of those blazing stars up there was once a planet where the monkeys started fooling around with uranium."

If the idea were true, I asked King, and if uranium really was supposed to be a catalyst for man's evolution, then why fight the inevitable?

"Western culture has this idea that development has to keep moving *forward*," she said. "I think we have to back away from that idea. Why is there this imperative to keep progressing? The hallmark of Western culture is to keep going and going and going. This is a compulsion, and an irrational compulsion."

She concluded, "I think we ought to leave it in the ground. We don't need it."

The friendship between the environmentalists and the Aboriginals was always shaky. Both sides accused the other of cynicism and of using the other to promote their own agendas.

Environmentalists grew frustrated with the slow decisions and arcane family feuds of the Aboriginals, who, in turn, sometimes felt like pawns in a media war. An Aboriginal woman named Jacqui Katone resigned from the Australian Conservation Foundation, saying she had been treated like "window dressing." And during the height of the Jabiluka protests, two young white men performed a ceremonial dance that they said was designed to raise the Rainbow Serpent in order to help the antiuranium cause. This was offensive to some of the Mirrar clan, who marched into the camp brandishing sticks and telling the protesters to quit their dancing and go the hell back to Sydney.

One thing these factions shared, however, was the Götterdämmerung view of uranium. Tony Grey knew this well, even though he was an ardent defender of extraction. "Its apocalyptic power, its lethal and invisible radioactivity, and its secrecy made it easy to demonize," he wrote. The Aboriginal vision of the serpent Dadbe, roused from his sleep and ready to destroy the world, had become a uniquely Australian metaphor for what uranium might accomplish once it was lodged in the warhead of an intercontinental ballistic missile.

The Aboriginals who signed the leases had made it known that their serpent's home on Mount Brockman was not to be touched under any circumstances. Today, there is a four-and-a-half-foot wooden fence that separates the mesa from the two large uranium pits. Any employee of the Ranger Mine found to have crossed the boundary without permission is subject to being fired on the spot. Sensors have been laid near the base of the mountain to record the seismic effects of the ammonium nitrate explosives in the pits, which are set off every other afternoon to jar loose the overburden.

I asked Joe Fisher about the holy places in the uranium country, and he screwed up his face.

"Most of the sacred sites are something they [the Aboriginals] made up as they went along," he told me. "The activists were just using the Aboriginals, trying to stop development, but it didn't work out that way."

He has a story about a company that started combing the Coronation Hill district in the early 1980s when gold prices were on the rise. Some Aboriginals were upset at the news. The gold target happened to lie under a place they said was an abandoned ceremonial site in a place known to them as Sickness Country, the home of a god called Nargorkun, who, like his relatives Dadbe and the Rainbow Serpent, could bring about the end of the world if he was provoked. Nargorkun grew sick, and so his two wives hunted food for him while he rested. If you happened to wander into the country without proper religious precautions, you could come down with the same wasting disease that had enfeebled the polygamist god. There were some etchings of him and his wives near a place called the Sickness Waterhole.

Years ago, while out prospecting, Fisher had discovered several examples of the Sickness art carved on rocks and had urged their protection from blasting. But he was convinced that none of those holy places was anywhere near Coronation Hill and made his views known to a Senate standing committee. An anthropologist was hired to investigate the matter and found that the Aboriginals who filed the complaint had only been shown *photographs* of Coronation Hill and had not made a visit themselves. The ceremonial site was eventually located at a spot thirty-one miles south, and the complainants admitted they had "made a big mistake."

"It makes me wonder," concluded Fisher, "how many other sacred sites have been proclaimed when they do not exist."

The incident underlined a touchy subject in the Australian uranium business. The gods leave no oracular evidence of their Dreamtime activities, and so "proving" the holiness of a piece of land is a highly subjective act. Mining companies have suspected overreaching, or even outright fraud, on the part of the Aboriginals. But Aboriginals have compared their earthen landmarks to cathedrals and wondered why these places historically were treated with such indifference by their white neighbors.

The British writer Bruce Chatwin spent several months in Australia in the early 1980s trying to learn a few Dreamtime stories. He remarked on the cleft between the European view of geography and that of the indigene, particularly in the face of bulldozers. "It was one thing to persuade a surveyor that a heap of boulders were the eggs of the Rainbow Snake or a lump of reddish sandstone was the liver of a speared kangaroo," wrote Chatwin. "It was something else to convince him that featureless stretch of gravel was the musical equivalent of Beethoven's *Opus 111*."

I went to see Yvonne Margarula, a senior member of the Mirrar, known for her willingness to talk to outsiders about the uranium mining. We sat at a picnic table outside the Gundjeihmi Aboriginal Corporation, a trailer office next to a mobile home resort, while some of her relatives' children played in the yard.

She was in her late forties, old enough to remember the time before the Ranger deposit had been discovered, and was of the last generation of Mirrar to have lived and hunted "bush tucker"—kangaroo, emu, grubs, and fruits—in the country near Mount Brockman. Fishing had consisted of throwing a sizable amount of eucalyptus bark into a pond, which temporarily deoxygenated the water. The suffocating fish floated to the surface and could be harvested with sticks. But this is not widely practiced today.

The number of full-blooded Mirrar has dwindled to twenty-six. They have taken on a seminomadic life at the fringes of Jabiru town; Yvonne's father, Toby, was said to have been mercilessly harassed, offered cars and booze in exchange for his signatures on various leases. The remaining members each receive $2,500 a month as a per capita payment for their bloodlines. None of them works at Ranger, despite multiple job offers. "They don't like it out there," one resident of Jabiru told me. "It's aesthetically unpleasant. There's acrimony. And they are unskilled."

My conversation with Yvonne Margarula was halting. What did she think of the uranium mining? She looked away from my eyes, the polite thing to do.

"Bad," she told me. "Mining is bad. We don't like it."

What about the leases? I asked.

"They gave us poisoned money," she said. "What is going to happen when they finish? They are gone, but we have to live here for life."

Why is the mining bad?

"Too many *balanda* around here."

This word, as it happens, is the nickname Aboriginals use to describe "whitefellows." The term is hundreds of years old and is not considered pejorative. It derives from the days when Makassarese fishermen from Dutch-controlled Indonesia visited and traded on the shores of northern Australia several generations before the British arrived in the eighteenth century. *Balanda* is a corruption of the word *Hollander*, but has become shorthand to describe anybody with white skin.

The term may not be racist, but what about the sentiment? "Too many *balanda* around here." If a white person started to complain about "too many blackfellows around here," wouldn't he be hounded out of the territory? I put this question to Graham Dewar, the director of the Gundjeihmi Aboriginal Corporation. He lit up a cigarette and offered the following exegesis.

"They were living an isolated life up until very recently. Thirty years of uranium mining is not that long. And so you have to remember the traditional view. The Western rational mind is in contrast to the blackfellows' way of thinking. Turning up in your neighbor's part of the country was always seen as an act of war. You were most likely there to steal wives or conduct sorcery or burn down the forest or other mischief. Unless you asked permission from the leaders to be there. In their minds, the uranium companies were invaders who never asked permission."

So, he said, when Margarula talks about "too many *balanda*," she is not talking about skin color, but about intruders who are trespassing.

"Whether they're here for uranium or coffee or beans, it doesn't make a difference at all," Dewar went on. "It just happened to be uranium. These are a bunch of assholes who don't care a bit about the people here and just wanted the pay dirt in the ground."

The Ranger Mine is scheduled to end its run by 2020. A large remediation plan is already being drafted, and ERA has promised to restore the countryside to an appearance similar to what was there when the first gamma rays were detected coming from the barge-shaped mesa.

"A lot of people say that when the mine closes down, the Mirrar will go back to living just as their grandparents did," Dewar told me. "Well, I can tell you that that's not going to happen. They have been too dependent on their incomes." Modernity, like uranium, had come to the jungle for good.

In addition to more than a dozen uranium sites left abandoned, there is a monument of sorts to Joe Fisher. A boulder affixed with a plaque was rolled next to a two-lane-highway bridge built over the Mary River in 1993. On September 1 of that year, a small party gathered around the boulder to dedicate the Joe Fisher Bridge. Fisher was there in the audience himself, a bit stooped, but his Walt Disney mustache was as black as ever.

Minister of Works Daryl Manzie made a brief speech. "In many respects, there are parallels between this bridge and responsible development, and the life and work of its namesake, Joe Fisher," he said. "Many would say that in Joe's case, mining and the environment seem an incompatible pair. But in reality, nothing could be further from the truth. Joe Fisher is living proof that you don't need to be an environmental vandal to champion the cause of the mining industry and, at the same time, promote sustainable development in the Territory."

Joe Fisher's marker stone has since been vandalized. Somebody pried off and stole the metal plaque, which had not been replaced as of the summer of 2007.

Australia continues to export nearly nine thousand tons of yellowcake every year, with nearly half that total coming from the Ranger Mine. This represents about 16 percent of world production. But there is plenty more for the taking. Nearly 40 percent of the world's untapped reserves of uranium are known to be located in Australia.

A leaching project called Honeymoon was on track to start operations in May 2008, adding a fourth uranium mine to the nation's roster. The then prime minister John Howard had signed an agreement with the Chinese government two years prior to supply twenty-two thousand tons of yellowcake per year to feed at least three dozen new nuclear power plants in mainland China. The Labour Party signaled its acquiescence by dropping the Three Mines policy. In the summer of 2007, when the spot price of yellowcake topped $120 per pound (more than four times what it had been in the days of the Uranium Club, after inflation adjustment), the hunt for new reserves was spreading across parts of the continent that had not been examined for decades, such as New South Wales, whose state government still categorizes uranium as a "contaminant" rather than a commodity.

"Australia has a clear responsibility to develop its uranium resources in a sustainable way," said Howard, "irrespective of whether or not we end up using nuclear power."

In many ways, the outlook for Australian uranium today is even brighter than it was in the frenzied years after Hiroshima, when Joe Fisher and thousands of others like him started combing the obscurity of the Dreamtime country, watching for the telltale jig of a needle.

Near the end of my afternoon visit with Fisher, I asked if he had any regrets about his long career.

"My only regret," he told me from his easy chair, "is that I didn't grab enough of the country. I would have claimed a lot more, if I could have."

7

INSTABILITY

To see the uranium in the African nation of Niger, you must take a bus from the capital city and ride more than twenty-four hours down a tattered ribbon of road that takes you to the edge of the Sahara Desert. The way is dotted with crumbling boulders and the occasional mud-walled village with sand piling up around low doorways and a few flayed goat carcasses hanging from poles, flies speckling the meat. Women with baskets full of onions walk barefoot alongside the fragments of asphalt that define the road, and men cover their noses and mouths with the tails of their head cloths as the bus chugs by. The houses have granaries to one side, constructed of mud and weeds and looking like swollen oriental teapots. Scattered acacia trees pockmark the desert, among patches of fibrous grass gnawed nearly to the roots by the camels and the sharp-boned goats.

Eventually the asphalt disappears, and the bus must find its way by following one of the multiple dirt tracks of the vehicles that have previously passed. When the bus bangs over a sinkhole or a rut, which is frequent, the coach rattles to its axles. Especially hard shocks tilt the bus to one side for a precarious second, lending the illusion, if not possibility, that it might capsize into the sand.

I had been on this ride for almost the entire day on March 1, 2007, sharing the bus with the members of a soccer team on their way to a tournament. The land had changed, from the dry to the drier, and now we were on a plateau, near the Air Massif Mountains, that looked like the surface of the moon. The road was somewhat better here, and it wound in lazy curves toward a high volcanic plain. The light was sunset-mild on the western hills, and the bus's long shadow slipped over the black rocks and sand on the edges of the roadbed. We were nearing a cluster of mud homes that a sign identified as Tagaza when the hijackers appeared.

A man stepped into the road brandishing a rifle and waved for our bus to pull over. Two other armed men were behind him, and three others, who had no visible weaponry, stood near a battered car that they seemed to be searching. They were apparently carrying out an ambush. I had my head down in a book and did not see these gunmen.

Our driver had the presence of mind to slam on his brakes and immediately throw the engine into reverse. The bus whined backward down the road for approximately half a mile. Out the window, I saw a group of children standing in the shade of a concrete hut, frantically waving their arms at us. *Get away, get away,* they seemed to be saying.

I was the only foreigner on the bus, but a member of the soccer team had spent some time in Nigeria and spoke a little English.

"What's going on?" I asked him.

"Didn't you see that?" he said. "We have to go back." He then described to me the scene that I had missed.

At this point, I was more annoyed than frightened. We had only a few more hours to go before we reached our destination—the uranium town of Arlit—and the ride had been tiring. I was eager to be done with it. The prospect of waiting several hours for the road to clear of bandits was not attractive. But I had no choice. The fifteen members of the soccer team, who had spent the last several hours laughing and joking, had gone silent. A few were ducked below their seats. I decided to keep my head down as well.

The driver continued his full-throttle reverse until he found a place in the sand sturdy enough to support a three-point turn. Then we were heading back down the way we came, following a line of power poles that marched toward a giant electrical plant. Lights winked midway on its smokestacks.

The sun had disappeared, and it was dark when we pulled up to a walled house in the town of Tchirozerine. This was one of the only villages in rural Niger to have electricity, and it was for the simple reason that it sat outside the gates of the French power plant that fed the nearby mines. Coal was burned here in order that uranium could be mined.

I stood outside the bus with the soccer team, kicking the dirt in a pool of blue security light, and we all watched as a jeep full of men in military fatigues pulled up and had a conversation with the driver. Then they pulled away.

I was made to understand that the soldiers were going to see if the road to Arlit was clear of the gang of bandits or terrorists—nobody was yet sure who they were—and that I, as a white man and an American, ought to have been especially grateful the driver did not attempt to plow on through as the armed men surely would have shot out the tires and boarded the bus. I then would have had special value as a hostage.

We were taken to the home of the mayor of Tchirozerine, a man who wore flowing purple robes and fed us three giant dishes of food that looked like spaghetti. The soccer team ate with their hands, squatting in front of the plates. They wore yellow and green jerseys with the legend AS. DOUANE and had been on their way to the biggest game of the year. But nobody seemed concerned about making it there on schedule. They had heard stories about people recently being kidnapped or killed in the desert. A rebel group had supposedly been active in the area, murdering soldiers before melting back into the desert, or over the border into nearby Algeria.

The mayor rolled out a carpet for us that seemed to be as big as a tennis court. We lay down on it in rows, breathing slowly, while mosquitoes feasted on us in the stale hot air. A few men snored, others murmured quietly in little groups. I tossed and dozed until dawn, when I crept out of the house to walk in circles around the mud-walled streets of the town until it was time to board the bus again at 8 a.m.

The mayor gave us all a little speech in French and waved at us with a smile.

"*Bon chance*," I heard him say. "Good luck."

We drove north for half an hour before coming up again on the culvert at Tagaza where we had nearly been hijacked. The soccer players sat upright and alert. Nobody spoke. From my seat in the middle of the bus, I looked down the culvert where the men had apparently been lying in wait the evening before. There was no sign of anyone, only a braided series of cattle paths that wound through the acacia trees and out of sight. Then we were back on the wide plain.

The city of Arlit announced itself with a huge mound of waste rock on the horizon. The terraced orange heap had been disgorged from an underground uranium mine called Akouta, which is the largest of its kind in the world and has been in production for nearly thirty years. Another mine nearby is a giant open pit. Both are controlled by a company based in Paris called Areva, and together they produce about 8 percent of the world's uranium. This is Niger's top export. The second is onions. There isn't much else.

As a consequence, Arlit is one of the only places where there seems to be a bit of money. There are wide dirt avenues, a few satellite dishes poking out from the roofs, young men clustering around parked Yamaha motorcycles, a few skeletal cell phone towers, and—exotic in this Muslim country—a small Catholic church tucked away on a side road. I talked with a thirty-one-year-old man named Abdoussalam who proudly showed me a picture of the 992G Caterpillar front-end loader he drove at the open-pit mine. It was the background picture on his cell phone screen.

"I have what I want, I have money," he said. "There is no sensation of danger."

There is a popular song here, a bouncy tune the lyrics of which are in the indigenous language of Hausa.

> *Miners, you struggle every day*
> *Miners, we should respect your money*
> *You earn it hard*
> *Miners, we salute you*

I talked on the phone with Abdoulaye Issa, the manager of the open-pit mine. He told me he was sorry to hear of the incident on the road. All of Niger's uranium production moves down the same route in convoys, about once a week. I asked him if there was any danger that a shipment might be stolen.

"It is well guarded and surveyed along the route," he told me. "We have been doing this for forty years. You can never be totally secure, but we take precautions to make sure that it gets to port in a safe way."

There is a mill at Arlit, and so the uranium leaves the city in the form of yellowcake, the pale grit also known as U_3O_8, which is the standard form for transporting the stuff over long distances. Anybody who stole a barrel would have a very difficult time using it for anything but decorative gravel. Less than 1 percent of it consists of U-235, the isotope that creates the blossoming chain reaction that makes fission and destruction possible. Yellowcake must be converted into uranium hexafluoride gas and run through an enrichment plant the size of a big-city airport before it can do any real damage. It is loosely packed in drums, loaded onto trucks, and hauled five hundred miles down a crumbling road known as the Uranium Highway. Despite its condition, this is regarded as one of the best roads in Niger. It terminates at the port of Cotonou in the neighboring country of Benin, where the ore is transferred to freighters and shipped to France. There it is stripped of useless U-238 and molded into fuel pellets to serve 80 percent of the electrical demands of the colonizing power. The street lamps in the old quarter of Rouen and the floodlights that bathe the Eiffel Tower are lit by uranium from Africa.

Though Niger is the fourth-largest producer of uranium in the world, it sees almost none of the wealth. Because of a long-standing contract, the French* consortium pays only 5.5 percent of its revenues in taxes, and most of it goes to subsidize elites in the dusty capital of Niamey. Almost three-quarters of the people cannot read, and those who survive to the age of forty-five are living on statistically borrowed time. Niger was recently named the most deprived country on earth by the United Nations, ranked dead last among the world's sovereign nations on a comprehensive scale called the Human Development Index, which charts life expectancy, education, and standard of living. Most people live in agricultural settlements and scratch out a meager income from onions, millet grasses, and goats. Irrigation projects are scarce, and so if the rains don't come, the people do

---

* Though Areva is the dominant player, smaller stakes in the mines are also held by corporate interests in Japan, Spain, and Niger.

not eat. A drought and an abundance of locusts ruined crops and killed livestock in 2005, causing a near-famine in the countryside. President Mamadou Tandja later complained that Western aid agencies had been exaggerating the drought, as well as Niger's dismal social rankings.

Unhappiness over hunger and the bleak future helped spark a rebellion called the Niger Movement for Justice, which started murdering soldiers near Arlit one month before I visited the country. The rebels are mainly Tuareg, a desert-roaming people descended from Berbers, who used to ferry salt by camel to the coast of the Mediterranean, and who had fought against French colonization. The Tuareg are proud and fierce fighters, sometimes called the blue men for the indigo-dyed scarves and turbans they wear for protection against the blowing sand. Never truly integrated into Nigerien society, they have traditionally been considered—and consider themselves—a people apart. The rest of Niger has not forgotten its history of trading slaves and raiding caravans and has been adamantly opposed to the creation of a separate nation-state. And yet, the violence is nothing new: Tuareg rebellions have been a feature of this part of the Sahara for more than a thousand years. Yellowcake mining now gives nuclear contours to these blood-grievances.

The Tuareg agitators were making themselves an embodiment of what the uranium business euphemistically calls geopolitical risk. On April 20, 2007, the movement raided a camp of uranium miners. They killed a guard and wounded three others before disappearing with six stolen vehicles and a number of cell phones. In June, they hijacked another bus on its way to Arlit and slaughtered three passengers. One of them was a two-year-old child. In July, a Chinese uranium geologist was kidnapped. The rebels released him after several days, but told all foreign mining interests to leave the area "for their own safety." For good measure, they sent twenty of their men to make an unsuccessful raid on the airport in the nearby city of Agadez and buried land mines along the Uranium Highway. A bus hit one of these mines in November and five passengers were wounded.

Their grievances echoed what the Tuareg had been complaining about for decades—corruption, racial discrimination, and unequal distribution of money from uranium mining. They also were angry about radioactive dust blowing onto their grazing fields, and the way that exploratory drills had started to show up on the plain where they held an annual salt cure rendezvous each September.

This was almost certainly the group that had attempted to hijack the bus that I had been riding in.

The spine of all of these attacks is the same highway on which trucks bear away the only tangible expression of wealth that the country has to offer, en route to its eventual consumption in French power plants. Niger's government has effectively lost control of this road.

The violence in the uranium fields is a classic outgrowth of what economists call the "resource curse," the unique misery laid upon those nations that sit atop a stockpile of a single desirable material—gold or rubber or lumber or (especially) oil. These nations ought to be prosperous, but are actually driven deeper into poverty. The natural treasure is locked up by a Western company, and whatever tax revenues there are are partially diverted to the president and his associates for their discretionary pleasure, leaving only scraps for the people. There is little incentive to develop a more healthy multilayered economy. If such a nation were a person, its diet would be of sugar and lard. Periodic insurrections force the government to use a heavy hand against troublemakers.

This is an old story in Africa, and Niger's uranium would be only one more example of a metal collar, except that a chain of bizarre events put it near the middle of one of the great foreign policy disasters of recent times. The uranium—or more precisely, the fear of it—would become a centerpiece of the American rationalization for invading Saddam Hussein's Iraq in March 2003.

The specter of uranium was something that H. G. Wells and William L. Laurence had understood perfectly. Now it was a royal road to war.

The uranium business in Niger was born in the mid-1970s, when the price of uranium was buoyant and the country was suffering from a crushing drought in which more than a million people died of malnutrition, starvation, and disease. Money was found, however, to construct a bold new headquarters for the Ministry of Mines, a building whose curved and glassy exterior bulges outward like a pregnant woman's skirt. A major street in the capital was rechristened Avenue de l'Uranium.

The president at that time, Seyni Kountché, had seized power in a military coup, and he was eager to use Niger's only dependable industry to boost the treasury. "We will sell uranium even to the devil if we have to," he said. This sentiment was mostly bluster, but it was remembered in Western intelligence circles.

On October 7, 2002, long after Kountché had been overthrown, a man named Rocco Martino took a woman out to lunch at the fashionable Bar Ungaro restaurant in Rome, which was down a short flight of steps from the street. Martino was an elegantly dressed man in his early sixties with gray hair and a thick mustache and something of the aspect of a faded lothario. He was there not to seduce, but to sell a scoop.

His companion was Elisabetta Burba, a reporter with the Italian magazine *Panorama*. She knew that Martino had contacts with the Italian intelligence agency, called Sismi, and he had previously sold her some newsworthy tidbits about peace talks in Kosovo and terror links at an Islamic charity. Such deals would be anathema in an American newsroom, but they are routine in the world of Italian journalism.

Over lunch, Martino asked her, "Do you know anything about the country that has sold uranium to Iraq?"

He handed Burba a file folder. One of the papers inside was a letter written in French, bearing the national seal of Niger and stamped with the words CONFIDENTIAL and URGENT. On the seal was the fuzzy-looking signature of the current president, Mamadou Tandja, and the papers appeared to confirm a secret deal under which Niger would agree to sell a massive amount of uranium to Iraq, apparently for use in that nation's attempt to build a nuclear weapon. Rendered in all capital letters, and in the style of the outdated telex machine, the communiqué announced that SAID PROVISION EQUALING 500 TONS OF PURE URANIUM PER YEAR WILL BE DELIVERED IN TWO PHASES. Martino also provided supplementary documentation, including memos from the Foreign Ministry and a photocopy of a twenty-five-year-old embassy codebook.

Where did he get such sensitive papers? "A source," was all he would say.

Martino wanted $12,000 for this information, according to accounts of the lunch published years later in the *Washington Post* and the *Wall Street Journal*. But all Burba could promise him was that she would try to validate their authenticity and get back to him. She related the exchange to her editor, who was a friend of Italian prime minister Silvio Berlusconi, himself a key ally of U.S. president George W. Bush. "Let's go to the Americans because they are focused on looking for weapons of mass destruction more than anyone else," said her editor. He told her to take Martino's documents to the U.S. embassy on the Via Veneto. She complied with the request and handed over copies of some of the documents—including the bombshell sales agreement—to the press attaché, Ian Kelly. The papers, later called "the Italian letter" by the journalists Peter Eisner and Knut Royce, were then forwarded to the State Department's Bureau of Intelligence and Research and into the chute of the U.S. intelligence network.

Their arrival in Washington that October coincided with a fervent campaign by Bush and Vice President Dick Cheney and other administration officials to sell the invasion of Iraq to the American electorate, and uranium was beginning to play a central role. Rocco Martino's lunch with Burba, as documented by Eisner and Royce, had been only one of the routes through which the tip reached the Bush administration. But this was apparently the first time anyone had seen the letter allegedly signed by Tandja. The CIA had been given a version of the same information by Sismi back in October 2001 and had been skeptical. The British government had also received the information through channels that have never been disclosed. On September 24, 2002, the British issued a dossier claiming (in the passive voice and without elaboration) that "uranium has been sought from Africa that has no civil nuclear application in Iraq."

This dossier served to fortify an emerging narrative. National Security Adviser Condoleezza Rice had appeared on CNN previously that month and conjured an image that would soon become a Bush administration mantra. "There will always be some uncertainty about how quickly [Saddam Hussein] can acquire nuclear weapons," she said. "But we don't want the smoking gun to be a mushroom cloud." The threat of sarin or smallpox was simply not frightening enough to justify an invasion. An invocation of the highest fear—nuclear apocalypse—was necessary.

Bush told an audience in Cincinnati, Ohio, on October 7, "If the Iraqi regime is able to produce, buy, or steal an amount of highly enriched uranium a little larger than a single softball, it could have a nuclear weapon in less than a year. And if we allow that to happen, a terrible line would be crossed. Saddam Hussein would be in a position to blackmail anyone who opposes his aggression....He would be in a position to threaten America."

The biggest applause came near the end of the speech, when the president declared, "We refuse to live in fear!"

One of the many doubtful aspects of these claims was that Iraq already had uranium mines in its desert interior, as well as 550 tons of yellowcake stored in warehouses inside the country. This was the legacy of a long-dormant nuclear program that had been shuttered in the early 1990s. There would have been no need to make such a risky and foolish deal with Niger.

There was also the problem of enrichment. Making yellowcake into the kind of "softball" described by Bush takes an industrial facility the size of a college campus. No such complex had been located by Western intelligence or the IAEA. Bush administration officials were then sent out to leak the dubious claim that Iraq had tried to build centrifuges, and as proof, they cited the interception of a shipment of sixty thousand high-strength aluminum tubes in Jordan that were almost certainly intended to be fashioned into surface-to-air missiles for conventional battlefield use.

These metal tubes were about the length of a baseball bat and had the circumference of a large grapefruit. U.S. Energy Department analysts considered them too small for the kind of rotations needed to separate the isotopes. The IAEA believed that they would have required elaborate retrofitting to be used to enrich uranium—their thickness ground down from three millimeters to one. Some of the tubes were even stamped ROCKET. The design perfectly matched the shafts of the same rockets that Iraq had used in its lengthy ground war with Iran in the 1980s. But in the *New York Times*, the lead paragraph of this story came out as "Iraq has stepped up its quest for nuclear weapons and has embarked on a worldwide hunt for materials to make an atomic bomb, Bush administration officials say."

Skeptics within the CIA and the State Department questioned the aluminum tubes story, as well as that of the African uranium sale, but the exhibits kept reappearing in the administration's public statements. Lawrence Wilkerson, the former chief of staff to Secretary of State Colin Powell, later complained that neoconservative officials close to Cheney had always insisted that the intelligence be taken at face value, despite grave internal doubts. His criticisms were later echoed by a host of ex-administration and intelligence officials who felt the uranium fears were being hyped. To their chagrin, the Italian letter had even found its way into the National Intelligence Estimate, an annual report that supposedly represented the very best information and analysis that the United States had to offer. It was now made to hint at midnight reagents simmering in the Mesopotamian desert—an apocalypse plot of The Other.

The selling of the Iraq war reached its apogee on January 28, 2003, less than three months before the war commenced, when Bush made his annual State of the Union address to Congress and uttered the now-notorious "sixteen words" that echoed the documents passed across the table of a Rome restaurant:

The British government has learned that Saddam Hussein recently sought significant quantities of uranium from Africa.

"Impossible!" This was the verdict of Moussa Souley, the local director of operations for Areva, the French company that has controlled nearly every aspect of Niger's uranium output for forty years.

I had been sitting with Souley inside his office near the Ministry of Mines and had asked him if there was any scenario, even a remote one, in which a foreign government or a terrorist group could have secretly bought yellowcake from the mills at Arlit.

"If somebody comes to us and says 'I want uranium,' they would have to go through the government. And then the government has to come and see us. This would mean we'd have to discontinue our existing contracts. And anytime we ship uranium there are multiple documents to sign. We know where it goes."

There would be, he went on, dozens of people who would have to be in on the conspiracy. Such a large shipment would have to be hauled by a flotilla of trucks, an operation that would attract widespread attention. Whenever uranium is transferred in Niger, there is a stack of paperwork that must be signed and stamped: bills of lading, bills of delivery, transport contracts, receipts, and tax documents. Concealing the paper trail would be a major undertaking. And it would be extremely unlikely that a French monopoly company would sell its radioactive product to a pariah state such as Iraq at the same time that its government was pushing the United Nations for more sanctions against the regime.

Souley showed me a graph depicting the historic output of the Akouta mine. It told the story of a long, dull marriage. The French company was extracting a reliable two thousand tons of yellowcake per year, every year, with almost no deviations.

"This is a conservative strategy," he said. "France likes the security of having its own mine, even though it would be cheaper to buy it on the open market." The letter furnished by Rocco Martino described a purchase of five hundred tons. That would amount to a quarter of the Akouta mine's annual production—a staggering pile of yellow grit the absence of which would have created immediate alarm in Paris and resulted in a scandal.

Souley summed up, "This kind of engagement would be visible by its nature. Somebody would see, and somebody would tell."

After she handed the Italian letter over to the U.S. embassy, the *Panorama* journalist Elisabetta Burba traveled to Niamey and reached much the same conclusion as Souley. Too many people would have been in the loop, and it made no sense that a pro-American government that depended on a healthy stream of foreign aid would have jeopardized its existence that way. Burba had already spent some time on the Internet looking up some of the names in the memos and had come up with disturbing inconsistencies. One cover document was signed by a foreign minister who had long since left that position. Another passage—laughably—was written in Italian. The fuzzy signature from President Tandja appeared to be a photocopied snippet glued to the page. Others noted that the letter made reference to the antiquated Niger constitution of 1966. The papers were forgeries, and not very good ones.

When she returned to Italy, Burba told her editor the papers were worthless and that Martino shouldn't get a cent for them. She offered to write a story on the deception, but it was never published.

Burba could not have known it then, but a retired U.S. diplomat named Joseph Wilson had visited Niamey earlier that year in a CIA-sponsored attempt to verify the story. He spent eight days at the Ganwye Hotel, the nicest in town, drinking mint tea at poolside with some of the people he had known in the government and at Areva from his days as a diplomat there in the 1970s. Wilson has an open face and a bearish charm, and the hotel staff took to calling him "Bill Clinton." He went back to Washington with the report that a midnight sale to Iraq, or anyone, was extremely unlikely.

"You're talking about a lot of trucks going north to south," he told me, years later, in a telephone interview. "How would you do this without anybody knowing?"

After the invasion of Iraq was over, and when it was becoming clear that "Saddam's nuclear program" had been a fantasy, Wilson made his African mission a matter of public knowledge in a *New York Times* op-ed. The administration sought to discredit him by leaking the news that his wife, Valerie Plame, was an undercover CIA agent who had recommended him for the task. The belief that such a fact would somehow invalidate his findings was exceeded in puerility only by the ham-fisted way in which the innuendo was fed to a series of Washington reporters. In the subsequent probe, Cheney's top aide, I. Lewis Libby, was convicted of lying to federal investigators.

The capital soap opera over Valerie Plame overshadowed a more essential question: Who forged the documents that Rocco Martino wanted to sell?

The answer seems forever lost in the swamp of Italian spy craft, though many theories have been advanced. It was neoconservative elements in the CIA who wanted to launder bad intelligence through the magazine *Panorama*. Or it was associates of Silvio Berlusconi, trying, as a favor, to give his American allies the excuse necessary for war. The journalist Craig Unger has speculated that the genteel paper peddler Rocco Martino was being used as a "cutout"—that is, an easily dismissed puppet—for the Italian intelligence agency, Sismi, which, though ridden with waste and incompetence, has known since the early cold war how to build mansions of smoke. Sismi itself has floated the story that the documents were a plant from the French government, which was eager to make Bush's atomic claims look ridiculous. Martino told several stories before refusing all further comment, but he has blamed a cabal within Sismi for perpetuating the hoax. Others have pointed to a middle-aged Italian woman working in the Nigerien embassy (known as La Signora by Italian prosecutors) who may have cut and pasted fragments of legitimate correspondence onto the documents, photocopied them, and then sold them to Martino, her only motivation being greed. Separate Italian and FBI investigations have yielded no solid answers.

"We may never know who forged them," Jacques Baute, the director of safeguards technology at the International Atomic Energy Agency, told me.

The IAEA was finally given an electronic copy of Rocco Martino's memo on February 11, five weeks before the Iraq invasion, and Baute, a courtly man with a trim white beard, had gone immediately to work. He told me it took him only fifteen minutes of investigation to conclude the papers were shoddy fakes and that the uranium deal was a fiction. A last-minute appeal to the United Nations could not stop the war.

Few things inspire the collective dread of the West as much as the suggestion that a poor country—particularly an Islamic one—is busy trying to acquire or enrich uranium. This apparition convinced the American public to support the dubious adventure in Iraq, and it continues to excite tensions with Iran, which has made no secret of its ambitions to enrich uranium.

The Iranian bomb project was apparently suspended after the 2003 invasion of neighboring Iraq, but its enrichment facility still exists. Capable of enriching as much as 165 pounds a day, the plant is hidden in the mountain town of Natanz, which is also known for its pear trees, its sharp cool air, and minarets that date to the thirteenth century. The highest local peak is named Vulture Mountain. A local legend says that a nearby valley was the spot where an invader from the West, Alexander the Great, slew the Persian king Darius III in battle in 330 B.C. But this is apocrypha. Darius was actually assassinated by one of his friends in a spot much farther away.

The enrichment facility is snugly underground, built as the Israelis had built Dimona, which makes it harder to see and harder to bomb. There are said to be three thousand centrifuges in a room the size of a professional basketball arena, capable of spinning uranium hexafluoride at speeds high enough to dislodge the U-235 atoms for collection. The centrifuges are copied from A. Q. Khan's basic model. Iran has also reportedly developed its own model of centrifuge, a type of design called IR-2, which is half the size and has four times the productivity. A bunker dug into a nearby hill is presumed to be a storage facility for centrifuge parts and the finished uranium.

Iran has danced around the question of why it needed to build Natanz, insisting that it has plans to erect three nuclear generating stations while at the same time refusing the idea of accepting uranium deliveries from elsewhere. It wants the means of production for itself. "We must not at the beginning of the twenty-first century revert to the logic of the dark ages and once again try to deny societies access to scientific and technological advances," said President Mahmoud Ahmadinejad in his maiden address before the United Nations in 2005. He later told a group of friends that a halo of light had been emanating from him as he spoke.

The former Israeli prime minister Binyamin Netanyahu once disparaged Ahmadinejad's government as "basically a messianic apocalyptic cult," and Iran's president has not entirely contradicted this impression.

Born the son of a blacksmith in 1956, Ahmadinejad holds a doctorate in "traffic and transport" and served as mayor of the capital city of Tehran before his election to the presidency in June 2005. He is slight in figure and soft in speech, and he prefers a tan jacket to a business suit. Ahmadinejad's closest allies are hard-line Shiite clerics, and he is said to have spent almost no money running for office, trusting in the turnout from the mosques to carry him.

He is also said to be a fervent believer in a Shiite folk belief: the return of the "hidden imam," a holy man who disappeared in the ninth century and is believed by Shiites to be the Madhi, a salvation figure whose dramatic reentry into the world will trigger a final confrontation between good and evil before the dawning of a final age of justice and peace. This is not found in the Koran, but millions believe it to be true. There is no set timetable for the messiah's arrival, though he is supposed to arrive with Jesus (regarded as a prophet in Islam, though not the Savior) and, in some versions, after a global war in which 80 percent of humanity dies.

In one of his first acts as president, Ahmadinejad approved a $17 million renovation of the magnificent blue mosque in the city of Jamkaran, where the Madhi is expected to reveal himself. There have also been reports that the president—a doctor of traffic—has studied the layout of Tehran to make sure the city can handle the crush of people who will arrive for the imam's first procession. Signs that said HE IS COMING went up all over the city after Ahmadinejad took office. "The prospect of such a man obtaining nuclear weapons," noted the London *Telegraph*, "is worrying." This is a thought frequently echoed by leaders in the United States and the European Union, who say they will never tolerate a nuclear Iran.

"We've got a leader in Iran who has announced that he wants to destroy Israel," said President Bush, adding, "[I]f you're interested in avoiding World War Three, it seems like you ought to be interested in preventing them from having the knowledge necessary for making a nuclear weapon."

This idea of "knowledge," and the way the West would deny it to newcomers, lies closer to the true motivations for the Iranian uranium project. Iran's nuclear thirst lies not so much in a desire to destroy the world, but rather in a yearning for lost prestige.

There was a time when Islam was the dominant faith in the civilized world, the center of a global empire bigger even than the Romans had made, and its denizens made bold strides in mathematics, astronomy, literature, architecture, and medicine. After the death of the Prophet, Muhammad, in the sixth century, his followers worked quickly to consolidate regional allies and spread the message that Allah was the One True God. Energy coursed through the movement: The willing received the message hungrily; the unwilling were put to the sword. Armies flooded out of the Arabian Peninsula and stretched the caliphate wide, punching into Spain to the west and India to the east and swallowing the lands where the patriarch Abraham had walked. The spiritual capital was established at the crossroads city of Baghdad, where the Abbasid dynasty took power in the eighth century. With a liberal and generous attitude toward the acquisition of worldly knowledge, the dynasty commenced what has been called the Golden Age of the Islamic empire. Jews and Christians, specially mentioned as "people of the book" in the Koran, were treated as privileged minorities and encouraged to contribute. The works of Euclid, Plato, and Aristotle were translated into Arabic onto scrolls made from a paper mill built according to Chinese specifications. Baghdad featured the world's first lending library and an advanced observatory. On the other side of the empire, the library at Córdoba had more than a half million volumes by the ninth century. By contrast, Europe's largest library, at the monastery at St. Gall in present-day Switzerland, held less than five hundred manuscripts.

The Arab genius Ibn al-Haytham conducted experiments with light rays and is credited with using an early version of the experimental scientific method to separate truth from error. Doctors invented the bone saw, forceps, and the clinical use of distilled alcohol (this last, ironically, an Arabic word). The greatest strides were in mathematics. The idea of the number zero was borrowed from Indian scholars and brought into standard calculations. The decimal system, quadratic equations, and the numerations that quickened everything were also popularized by the Arabs. For the world at the time, this was a high-water mark of science.

Then it all changed. The Mongols sacked the intellectual center of Baghdad in the thirteenth century and, while political power lived on inside the vast Ottoman Empire, the spirit of learning and discovery never really recovered. The advances of Renaissance Europe were ignored. The printing press—that great democratizing force—was generally not welcomed.

"In the Muslim world, independent inquiry virtually came to an end, and science was for the most part reduced to the veneration of a corpus of approved knowledge," wrote Bernard Lewis in *What Went Wrong?*, an autopsy of the Golden Age. The influential sect of Wahabbism, founded in the 1700s, condemned the authority of science. After the disintegration of the Ottoman regime in the 1910s, Islamic societies turned ever more inward, shunning the Western idea of scientific progress as being counter-

Koranic. The Moroccan sociologist Fatima Mernissi has written, bitterly, that "Islam is probably the only monotheistic religion in which scholarly exploration is systematically discouraged, if not forbidden." Patents granted inside Muslim countries lag far behind those in other parts of the world, and top Pakistani physicist Pervez Hoodbhoy recently characterized the universities in his country as "intellectual rubble," with barely five qualified mathematicians among them. He noted the disquieting fact that Spanish publishers now translate as many books in a single year as Arab publishers have translated in the last twelve hundred years since the reign of the caliph Mamoun in the ninth century. The West is resented for its political and cultural dominance at the same time that many of the technologies that made it possible are disdained and rejected.

But the atomic bomb is the great exception to this rule.

The political decay of the nineteenth century and the oil colonization of the Islamic heartlands in the twentieth century have been inglorious, and some political leaders have wondered if there might not be a way to create a fast solution to the trouble—a magical way to catch up. The quest to attain nuclear capability is a matter of especial pride among hard-line factions in Iran who see it as a route to the reborn glory of the Islamic empire and—on a smaller scale—a way of igniting a national burst of confidence such as came to Pakistan when A. Q. Khan succeeded in making that country a member of the nuclear club.

"The bomb looms large in the popular Muslim consciousness as a symbol of Islamic unity, determination and self-respect," wrote Hoodbhoy. "It is seen by many as a guarantee against further humiliating defeats, as the sign of a reversal of fortunes, and as a panacea for the ills that have plagued Muslims since the end of the Golden Age of Islam. Such sentiments are echoed by Muslims from Algeria to Syria and from Iraq to Pakistan."

Even as antediluvian a figure as Osama bin Laden—while hardly a friend of the Shiite government in Tehran—has promoted atomic science as a means of Islamic advancement, if only as a sledgehammer to use against infidels. He has called it a "religious duty" for Muslims to acquire nuclear weapons in defense against the West. "It's easy to kill more people with uranium," one of his followers has said. Three years before the September 11 attacks, bin Laden put out a directive titled "The Nuclear Bomb of Islam," telling his readers that it was their obligation "to prepare as much force as possible to terrorize the enemies of God."

His own attempts to make or purchase a bomb have been clumsy. In 1993, a Sudanese military officer who called himself Basheer left bin Laden the victim of a $1.5 million con job. He sold bin Laden a tube of "uranium oxide" that turned out to be red mercury, a useless powder.

The father of the Iranian nuclear program is Akbar Hashemi Rafsanjani, a wealthy pistachio farmer and former military chief known as the Gray Eminence. At the end of the pointless Iran-Iraq war in 1988, he told the government disgustedly that international laws "are only drops of ink on paper," and that the nation would have to go beyond conventional weapons to achieve security. That same year, Iran received a block of uranium from the then apartheid nation of South Africa and began to experiment with it, along with some plutonium from a five-megawatt research reactor in Tehran that had been supplied by the United States, through the Atoms for Peace program (a legacy of better times). A. Q. Khan was paid $3 million for a set of technical specifications and centrifuge parts. A good domestic source of uranium ore was discovered in the Great Salt Desert, and even more was quietly purchased from China. But Iran has had difficulty finding qualified scientists to run the program.

Rafsanjani has made repeated calls for his nation to become more scientifically literate. His speeches often conflate the idea of "science" with "uranium enrichment." The two seem to have merged in the minds of the Iranian leadership.

"The natural right of a country which wants to make use of the latest sciences is under assault," Rafsanjani told a group of students in 2006. "The root cause of these assaults lies in the colonialist nature and policies of the West, whose plan is to keep countries backward." To another audience, he said, "Unfortunately, the world of Islam is in need of Western science. The Islamic revolution is determined to return that glorious era to the world of Islam. That is why the enemies of Islam are hurling obstacles under different pretexts." A nuclear war with Israel would leave Muslims the clear winners, he has reasoned, because a single explosion over Tel Aviv would decapitate that country, whereas the belt of Islamic countries from West Africa to Indonesia would absorb partial damage at worst.

This narrative of Islamic triumphalism often creeps into any discussion of uranium inside Iran. The religious scholar Karen Armstrong has noted that Islam places a premium on worldly results, in contrast to Christianity, which tends to see its highest expression within the context of visible *failure*—poverty, mortification, and crucifixion. Islam prefers more tangible evidence of divine favor, and the lack of modern atomic potency has been particularly crushing in this regard. The nation's supreme spiritual leader, the Ayatollah Ali Khamenei, has called Iran the "mother of science," who now deserves the "sweet fruits" of nuclear power. "At this time, God wants us to make what we need," one prayer leader said in a sermon attended by a Western reporter. "Other countries now feel threatened because we have advanced in our technology."

Rhetoric from the top has filtered to the street. While many average Iranians are unhappy with the way their leadership has taunted the rest of the world, there are others who take to heart the slogan that blares out from signs and is constantly repeated on television: "Nuclear energy is our indisputable right."

The idea has become so glorified in Iran, even fetishized, that it seems to cover up no end of other internal shortcomings. The righteous struggle to make uranium is a means for Ahmadinejad to blame the West for his nation's troubles, to direct the national anger outward instead of letting it focus on his own inabilities. This might be called the William L. Laurence view—uranium as messiah—with scant regard for the more banal and disappointing reality that it usually brings.

No matter: The goal of becoming a uranium maker has become a national battle standard, "an emotive nationalistic issue for Iranians, like supporting their football team," in the words of one political science professor at the University of Tehran. He was talking to a U.S. journalist, who also picked up this telling comment from a young Iranian woman: "For a country to have nuclear energy means that it has made progress in all other fields as well, so other countries have to respect its technology."

Iran has repeatedly insisted that it wishes only to have control of the fuel supply for the three reactors it plans to build and does not want to "lose" its uranium the way it gave up its oil to British companies in the 1910s (Niger might be said to have lost its uranium to the French in a similar way). There is no plan to build weapons, says Ahmadinejad, who also says such a thing would be against the dictates of the Koran. But even a beginning nuclear engineer knows that the cascades designed to produce 3-percent uranium (for power) can yield 90-percent uranium (for weapons) with a few metaphorical twists of the wrench. The same car that gets you to the grocery store can also take you to the ocean if you point it in that direction and drive long enough. This is uranium's joke on man: its refusal to be encircled; part of the "sheer cussedness of nature" that Enrico Fermi noted during the Manhattan Project. In fact, blending uranium up to weapons grade becomes physically easier and takes less time once the threshold of 3 percent has been crossed.

There are many reasons—even logical ones—for Iran to desire to cross this line. For one thing, it already lives in a nuclear neighborhood. To the north is Russia, east is Pakistan, west is American-occupied Iraq, and then Israel on the Mediterranean. Having a bomb of its own would allow for some rough parity and strengthen Iran's hand in the region. Once banked, the weapons would also reduce the chances of its leadership being the target of "regime change" by an invading superpower, or of another territorial conflict such as the one fought with Iraq in the 1980s, which served no purpose and wasted the lives of millions. This is the same philosophy of deterrence embraced by Bernard Brodie and the other U.S. strategists of the cold war, and what led to America's $10 trillion expenditure during the arms race (uranium is a costly servant).

But there is also the crude schoolyard calculus of international affairs to consider. Unfortunately, a weaponized state enjoys a level of prestige unmatched by lesser nations, though it may have turned itself into a pariah to get there. To join the elite circle of the nuclear club, even through the backdoor, is still a way of belonging.

The Iranian thirst for atomic potency approaches the level of a national fetish; a state of mind the psychiatrist Robert Jay Lifton has called "nuclearism," in which the power to shatter the atom and fry the enemy is worshipped as a thing it itself. The bomb overtakes all other considerations and blots out all alternatives. Lifton writes: "It is the ultimate paradox in human existence—the worship of the agent of our partial annihilation. It is not surprising that the weapons should become agents of worship because they could do what only God could do before, i.e., destroy the world."

He also says this: "Indeed, nuclearism can become sufficiently perverse to reach the point of seeking an experience of transcendence via a final nuclear apocalypse—which is something on the order of the sexual perversion in which orgasm is sought at the point of death via strangulation or hanging."

But it is not only Iran that is making a fetish of its uranium-making knowledge. Keeping that capability away from Ahmadinejad at all costs has become a priority of the United States, which has put pressure on Russia, China, and European allies to isolate the regime and keep it from joining the world's most select club.

Tehran clearly understands the apocalyptic pull of uranium, and it might also understand the singular effect that it has on the Western mind; the dread that our own hideous discovery could be used against our regional friends or, with the help of a missile, Miami or San Francisco or, God forbid, our own homes and children; the original sin of Hiroshima rendered back to us in a burst of savage white light. Such a terrible potency, it is thought, should never be trusted in the hands of The Other, the barbaric people on the other side of the hills: those who would take our lands, rape our brides, and slay our children if we are not evermore vigilant.

Man's most carnal tendencies are inflamed by the most modern of elements, uranium.

Iran has no monopoly on this dread, of course. There is no shortage of people in the world with grudges and visions, and an underground group that was determined to bring a final orgiastic reckoning would likely avoid using plutonium as a tool. Making it requires a nuclear reactor, for one thing, which calls for either the cooperation of a state or the help of several coconspirators within a facility. These places watch their plutonium like a miser counts his gold, and so an elaborate bookkeeping fraud would be necessary. The metal is also extremely radioactive. Without costly precautions and a shielded glove box, it would be likely to kill anyone who tried to fashion it into an implosion bomb. One thousandth of a gram of plutonium, if inhaled, causes death in a matter of hours.

Uranium is the far better choice. It emits only a lukewarm level of radiation. With a thin lead sleeve, it can be smuggled through those few border checkpoints that are equipped with working radiation detectors. It can also be molded into the same pellet-and-cup shape of the Hiroshima bomb with relative ease and for a low cost. The core of the first nuclear weapon built in China was lathed by a single technician in one night with ordinary machine-shop equipment. This was more than forty years ago, and the physics have not changed, even as the tools can now be ordered over the Internet. The uranium, once acquired, can be melted and shaped mostly by gadgets procured through Home Depot, with little or no danger to the craftsman.

"It's sadly not difficult," said the weapons expert Ashton B. Carter before a U.S. Senate committee in 2006. "You know that the United States had no doubt that ours would work, our very first one. . . . Any knuckle-head who has enough highly enriched uranium can make it go off."

The craftsman would be only one member of the team. There would have to be at least one physicist knowledgeable of the exact designs, as well as an explosives expert to gauge the correct speed of the uranium bullet fired into the core. A 1977 study by the U.S. Office of Technology Assessment said an atomic bomb could be produced by just two determined experts for a cost that in today's dollars would be about $3 million, accounting for salaries and materials costs. The Harvard professors Peter Zimmerman and Jeffrey Lewis have written that a more realistic bomb staff would consist of about nineteen members and have a budget of $10 million. The finished weapon could be hidden in an enclosed truck or a shipping container and taken to its target under the cover of normal harbor or highway traffic.

This is the easy part, as the "secret" to constructing an atomic bomb has been more or less public knowledge since the end of World War II. The hard part is finding enough uranium to make a core. As one expert put it to me, "The biggest challenge is not the design. It's the material." He was talking about a chunk of uranium roughly the size of a football. It must be highly enriched—stripped of almost everything but the angry atoms of U-235—and such cargo is best acquired from the shell of a decommissioned weapon, most likely from the former Soviet Union.

Making it independently is out of the question. That would require an industrial effort on the level of Oak Ridge. This bootleg uranium won't be coming from a nuclear power station, as the uranium would likely be in the form of pellets and fuel rods that already have been radiated and would be fatal to their handlers. Stealing the highly enriched variety from a weapons stockpile is "by far the most direct shortcut for actors seeking nuclear explosive capabilities," concluded two Swedish experts in a 2005 briefing.

The stuff has gone missing many times. In 1951, three boys in the prairie town of Dalhart, Texas, discovered a black rock lying near the railroad tracks. It was weirdly heavy—thirty pounds—though only about the size of a hamburger. The boys found that it made colorful sparks when they pounded on it with a hammer. The editor of the local newspaper believed it might be a meteor and sent it off to the University of New Mexico for testing. The rock turned out to be highly enriched uranium, apparently stolen from the laboratory at Los Alamos. An even bigger chunk of it was discovered in a nearby junkyard. If slammed together correctly, these two pieces would have leveled everything within ten miles. How they found their way to the Texas panhandle was never disclosed. The year before, a research scientist named Sanford Simons was arrested in Denver after the FBI found a glass vial of plutonium and several pieces of uranium tucked inside a dresser in his suburban home. "I just walked out with it," he told a newspaper. He said he just wanted "a souvenir" of his work, but he served eighteen months in jail for the stunt.

Those who work in the nuclear business are understandably sensitive about such incidents, which are classified under the rubric of MUF, or "materials unaccounted for." Uranium is transported in many forms— raw ore, yellowcake, hexafluoride, metal oxide, ceramic pellets, fuel rod assemblies—and at every step there is potential for carelessness. In 1969, a bottle of enriched uranium gas went missing and sat ignored in a freight storage room at Boston's Logan Airport before somebody finally tracked it down. The Nuclear Fuel Services Corporation's plant at Erwin, Tennessee, may have set the record for the sloppiest oversight of fissile materials. It admitted in 1979 that it could not locate up to about 20 pounds of highly enriched uranium, and a later investigation determined that as much as 246 pounds of uranium and plutonium had gone missing over the years. It may have been caught in the network of pipes and tanks like so much crusty residue, or leaked away as a gas. It may also have never existed at all, the ghost result of miscalculations about exactly how much was passing through the enrichment cascade. Or it may have been stolen by an insider using the "salami trick" of many tiny pilferages that add up to a significant theft. Operators at the Erwin plant, which fabricated submarine fuel for the U.S. Navy, were unable to explain the loss.

When measured against such incidents, the theft of fuel rods in the Democratic Republic of the Congo starts to look less like corruption and more like a routine happening. Indeed, the trade in stolen uranium is almost as old as the nuclear age. Just one year after Hiroshima, U.S. intelligence agents managed to infiltrate a speculative black market in uranium among several merchants in Shanghai. A nearby mine was rich in uranium crystals, and local entrepreneurs—mindful of recent headlines—were selling them for $5 a pound. More recently, the Revolutionary Armed Forces of Colombia was caught with sixty-six pounds of illicit uranium after a raid in March 2008. Experts said the guerrilla organization was not trying to build a bomb, but merely hoping to sell the uranium for a profit.

The closest thing to a police blotter the world has for incidents such as this is kept at the headquarters of the International Atomic Energy Agency, an ugly collection of Y-shape buildings next to the Danube River on the eastern edge of Vienna, Austria. The United Nations located several of its ancillary offices there in the 1970s because the real estate was cheaper than in Switzerland. The buildings were teeming with asbestos and recently had to be stripped to the pillars. They also now harbor a number of electronic listening devices. It is widely assumed that certain offices are bugged by member states looking for information about the nuclear capabilities of rivals; the espionage is said to get especially thick during negotiations.

In the squeaky-floored lobby, golf shirts and key rings are for sale bearing the legend ATOMS FOR PEACE, a coinage of Dwight Eisenhower's that reflects the somewhat schizophrenic mission of the agency. The IAEA is charged with promoting nuclear power throughout the world, while at the same time it is responsible for detecting the unauthorized uses of uranium and plutonium, and untangling riddles such as Rocco Martino's Italian Letter and Iran's level of technical prowess at Natanz. And all of this is supposed to be done (as its staff is fond of pointing out) on an annual budget that is less than that of the Vienna police department.

I went to the IAEA on a mild January day to see Richard Hoskins, an American who manages the agency's Illicit Trafficking Database, a compendium of the sixteen known incidents since 1993 that involve stolen quantities of plutonium or uranium, especially the bomb-grade variety known as highly enriched uranium, or HEU in the parlance of weapons inspectors. Hoskins's job title is senior staff member in the Office of Physical Protection and Material Safety in the Safeguards department. He told me that the sixteen incidents certainly represent an undercount because the agency has no investigatory capabilities of its own. It must rely on the willingness of its member nations to pass along reports from their own law enforcement agencies, and they are not always willing to do so. For one thing, police are genetically bred to hoard information. The admission of a nuclear crime may also be regarded as a diplomatic embarrassment to a member state.

"They are under no obligation to report anything," Hoskins told me. "We can cajole and plead and do everything short of offering money. It may reach us in days or months or years." While explaining this limitation to me, another IAEA official quoted Stalin's dismissive remark about the power of the Catholic Church: "How many divisions does the pope have?"

I asked Hoskins how much of the smuggling was going undetected, and he said the best estimates put it at about 80 to 85 percent—roughly on par with the percentage of cocaine that makes it through international waters or airspace on the way into the United States. "One of the worries we have is that we're only catching the dumb guys, and the smart guys aren't getting caught at the border," he told me. "Organized crime will do things a lot better than the amateurs we pick up."

Reading through Hoskins's list of incidents brings home the essential bathos of the exercise. Those who seek to discover and prevent crimes with uranium are in the unenviable position of trying to prove a negative. The promiscuous spread of uranium during the cold war means that nobody will ever be able to prove conclusively that a group somewhere does not have its hands on some for an apocalypse project. There is simply too much uranium around. Too much is out of sight. The list is a peephole at best, rendered in passive language; brief headlines with no narrative so as not to offend the member state giving up the information, and so each incident is like a dot. The effect is therefore impressionistic: May 10, 1995—"Tengen-Wiechs, Germany: Plutonium was detected in a building during a police search." March and April 2005—"New Jersey, U.S.A.: A package containing 3.3 grams of HEU was reported lost." December 14, 1994—"Prague, Czech Republic: HEU was seized by police in Prague. The material was intended for illegal sale."

Something is known of this last affair. The uranium had been stored in two metal canisters in the backseat of a dark blue Volvo limousine parked on the street. Czech police acting on a tip confiscated the powdered uranium dioxide and arrested three men, one of whom was a nuclear physicist. The other two carried Russian passports. The uranium was presumed to have come from an enrichment facility in the former Soviet Union that made rods for navy submarines.

Though there has never been a complete accounting, there is known to be at least one thousand tons of enriched uranium stored in Russia, the fruits of Wismut and St. Joachimsthal—an amount equivalent to at least sixty thousand nuclear warheads. Much of it is located in warehouses near the once-secret city of Ozersk in the Ural Mountains, a place that functioned as the heart of Russia's own Manhattan Project. During Boris Yeltsin's freewheeling and anarchic presidency in the 1990s, the U.S. government paid for a secure disposal site in the city nicknamed the "Plutonium Palace" to house at least some of it.

The uranium cores of old Soviet weapons were pulled apart and shipped to the United States for conversion into a lower-enriched form that will provide 20 percent of the nation's electricity needs through 2013. This was called the Megatons to Megawatts program, and the glut it created sent the global uranium trade into a decade-long depression. Every time you turn on the lights in America, there is a one in ten chance that the power is coming from an old Soviet warhead. Some of this material was destroyed. An official from the U.S. Department of Energy told me he witnessed blocks of highly enriched uranium set afire in a shielded chamber in a laboratory in Siberia. "I've stood in front of the glove boxes and warmed my hands on it," he said. "It's very beautiful to watch it burn."

Yet nobody is sure how much of the remaining material is under competent guard and how much has vanished in the nearly two decades since the fall of the Soviet Union. Though the operations are better managed than they were in the wild days immediately after the collapse of the USSR, there are still reports of shoddy security behind the cordon at Ozersk—bomb fuel stored in plastic buckets, flimsy wire fences, and guards who like to drink on the job. The Harvard researcher Matthew Bunn, who produces an annual report called *Securing the Bomb*, cited a persistent lack of funding and attention from the Russian government toward locking down old weapons fuel, despite the obvious high stakes involved.

"Both Russian and American experts have reported a systematic problem of inadequate security culture at many sites—intrusion detectors turned off when the guards get annoyed by their false alarms, security doors left open, senior managers allowed to bypass security systems, effective procedures for operating the new security and accounting systems either not written or not followed, and the like," he wrote. "The security chief at Seversk, a massive plutonium and HEU processing facility, reported that guards at his site routinely patrolled with no ammunition in their guns and had little understanding of the importance of what they were guarding."

Stopping a burglary at these facilities represents the very best chance for stopping a terrorist group from wiping out an American city, just as kidnapping victims stand the best chance of escape in the first three minutes of their abduction. But getting away with a backpack full of uranium would be only the starting point. A thief would have a lot more work ahead before he would see a dollar of profit. Whoever managed to get a block of uranium out of this place would have to first move it out of Russia, and the most convenient route would be down one of the great smuggling thoroughfares in modern Asia, through a pass in the heart of the Caucasus Mountains and into the broken nation of Georgia.

In the video, the smuggler looks calm. He wears a skinny mustache and a black leather jacket and keeps a stone face after the men he has been negotiating with in the living room have shown him their badges and told him he is under arrest. He obligingly reaches into the pocket of his jacket and pulls out a plastic zip-top bag full of dark powder that looks like instant coffee crystals. Then he shakes it gently back and forth, as casually as if it were a bag of potato chips, to show the undercover officers what he has brought for them. Oleg Khinsagov, fifty-one, had been caught in the act of trying to sell four ounces of highly enriched uranium, with the promise that he could furnish lots more of it, hidden in his apartment back across the border in Russia.

The meeting in the shabby tenement in the city of Tbilisi had been a setup all along, and a tactical police unit had been waiting outside the door in case Khinsagov had tried to run into the bathroom to flush the uranium down the toilet. The meeting in January 2006 had been a sting operation by the Interior Ministry of Georgia, which had been prodded by U.S. intelligence agencies to do more to catch smugglers offering old Soviet nuclear goods along with the usual cheap food and drugs. A detective with olive skin who spoke Turkish had been spreading the rumor among Georgia's gangster class that he represented a "serious group" of people in the Middle East who would like to acquire some uranium suitable for making a bomb. Khinsagov eventually surfaced and offered to sell him a sample packet for $1 million.

He looked just like any other weary workingman from an out-of-the-way Russian town. His usual occupation, he said, was exporting fish and sausages. He had also driven tractors and repaired cars. A previous career as an oil-field worker had taken him to rigs in Iraq and Dubai. He initially tried to claim the uranium deal was a scam of his own making and that the dark powder was just ground-up ink from a computer printer cartridge. An American scientific analysis said otherwise. The stuff was uranium enriched to nearly 90 percent, the perfect blend for a bomb. This could have come only from the reserve of a superpower.

Khinsagov displayed a surprising cockiness once he was in custody. "Usually if you're going to prison and you have eight years ahead of you, you try to be nice and make friends with your interrogators," said Shoto Utiashvili, the head of the Department of Analysis in the Interior Ministry. "But this guy was so arrogant. He tried to show us that he was standing firm. He was offered many benefits, but he never named the source. This is not logical behavior for a low-level smuggler. The logical thing you can deduce is that he was promised protection in exchange for his non-cooperation with us. Most likely, he was a middleman."

Georgian police never learned who supplied him with the uranium, or whether he had been telling the truth about the 4.4-pound cache supposedly back in his apartment. If he had been as good as his word, there would have been enough uranium for a device big enough to destroy a good portion of downtown Chicago. Before the undercover officers revealed themselves, Khinsagov had bragged that he had a friend who worked as a security guard who could help him move the larger package across the border. But Russian authorities refused to cooperate with the investigation and displayed little interest in finding out where the uranium had been processed. Khinsagov eventually stopped talking altogether and was convicted in a closed trial of "possession of a hazardous material" and sentenced to eight and a half years in prison.

His case resembled that of Garick Dadayan, a petty smuggler who had been arrested three years prior at the Armenian border, carrying a tea box with six ounces of highly enriched uranium. He told police he had been hired to deliver the package to "a Muslim man."

There is no way to know, Utiashvili told me, how many other uranium salesmen may still be operating inside Georgia. With a history of corruption and a long porous border with its disliked Russian neighbor to the north, the former Soviet republic is one of the most likely spots on the globe to serve as a transfer point for garage-sale uranium on its way to a terrorist group. The whole country is a bit smaller than South Carolina, but within its borders there are two breakaway "nations" where the central government has no military or judicial control and where the smuggling of food, gasoline, vodka, cigarettes, drugs, and other goods from Russia and back is a routine affair.

One shibboleth among weapons experts is that new patterns of atomic smuggling are most likely to map themselves on top of more established corridors where law enforcement is weak or nonexistent or at least bribable. Soldiers on the Russian border are known to ignore particular cargoes, for a price.

"It starts simply, with cigarettes and oranges," said Irakli Sesiashvili, the head of a military watchdog group. "Then comes the people who will set up deals for more interesting products, such as drugs, guns, or uranium. It's easy to find these guys who will work these deals within the military."

The culture of fraud does provide one safeguard. A Georgian police official told me that as many as 95 percent of the smugglers who say they have centrifuges or uranium turn out to be con artists themselves who sell only worthless chemicals or broken equipment. Red mercury, also known as cinnabar, is one common fill product. There have been cases, too, where the seized uranium is far below bomb-grade enrichment (most likely stolen from a power plant) and is also therefore useless. Helmet-shaped casings from reactors are said to be especially good for nuclear rip-offs because they have radioactivity symbols convincingly stamped upon them, though they contain nothing of value. Those who deal in such goods would be the spiritual cousins of the nervy Sudanese individual who conned Osama bin Laden out of $1.5 million for phony uranium.

Georgia has become an atomic crossroads at least in part because of its geography, some of the most spectacular on the Eurasian landmass. A popular folk myth says that the people of Georgia were having a binge at the time of Creation when God was busy handing out land to all the peoples of the earth. The Georgians arrived late and hungover. God told them he was sorry, but that no further territory was available. The Georgians thought quickly and told him they had been late only because they were raising their wineglasses in toasts to God. This pleased the Almighty, and he gave them the most beautiful land in the whole world, which he had been saving. The nation was thereby founded on a scam.

Perhaps there is a shade of knowing guilt to this story, as Georgia's spectacular physical setting has also been its curse. For centuries it served as a buffer between Mother Russia and the Islamic principalities to the south. Successive waves of Byzantines, Mongols, Ottomans, and Cossacks have ransacked their way through the mountain passes on their way to greater glories. The capital city of Tbilisi has been destroyed and reconstructed an estimated twenty-nine times, by one count. Many of the Orthodox monasteries resemble hilltop fortresses. The people here—not ethnically Slavic—nevertheless preserved their own language and poetry in the midst of the periodic sackings and conquests. The region first came under the knout of the Russians under Tsar Paul I in 1800 and was later forced under Communist rule in the 1920s. A onetime Georgian seminary student who took the name Josef Stalin (for "man of steel") maneuvered and backstabbed his way to the head of the Communist Party, a source of considerable embarrassment in Georgia today. When Stalin came to power, Georgia was swallowed up into the USSR and became a favorite warm-weather getaway for the elite. It also became the site of the largest metal fabrication plant in the world. A new city named Rustavi, a grid of exquisite awfulness, was built to accommodate the labor: a phalanx of gray apartment blocks marching in relentless sequence toward the soot-blackened mills.

Georgia won its independence in 1991 and immediately drew close to the United States, offering easements for a gigantic oil pipeline from the Caspian Sea to Turkey and stripping all signs with Russian letters from the roadsides and public buildings. But under the haphazard leadership of former Soviet foreign minister Eduard Shevardnadze, the nation fell into a state of kleptocracy. For nearly a decade, the famous Article Fifteen of the Congo appeared to be the only relevant statute on the books. Georgia became a place where literally everything was for sale.

"The police disappeared, the courts disappeared, the government disappeared. There were only people with guns," said Alexandre Kukhianidze, a former professor of political science who now runs a nonprofit group. "There were only two to three hours of electricity each day. Ministries turned into pyramid schemes."

As much as 60 percent of the cash flow in those years was located in a parallel market of bribes, smuggled goods, and counterfeiting. Western foreign aid also disappeared into the sinkhole. The regions of South Ossetia and Abkhazia formed rump provisional governments and expelled federal troops. At some point in the chaos, about four and a half pounds of highly enriched uranium disappeared from a laboratory shelf in a technical institute in Abkhazia. It has never been located. Shevardnadze was ousted in the bloodless Rose Revolution in 2003, and most of the police force was fired in the subsequent reform. Five years later, tensions over South Ossetia created a vicious shooting war with Russia, which Georgia accused of trying to undermine its territorial integrity.

Tbilisi is today a city of brutalist Soviet tenements with bunches of colorful plastic flowers draped over the balconies and oppressive concrete superblocks with vodka bars and young women in spike-heeled boots lighting cigarettes outside sad little shops without electricity that sell fatty sausages and Mars bars. Across the street from the opulent Parliament building is a kiosk shielding a hardy survivor: A rotary-dial pay telephone from the 1960s that still emits a dial tone. Soldiers wear uniforms with Velcro strips for the red-and-white Georgian flag pressed on their shoulders. The United States has invested heavily in the security infrastructure here—it sees a valuable regional ally, as well as a host for the Caspian pipeline project—and has built a series of new checkpoint stations with radiation detectors. One of them on the Armenian border cost $2.4 million in a grant from the Department of Homeland Security. The FBI has helped train some of the guards.

"To take contraband through our territory takes a little more effort now," said the deputy minister of defense, Batu Kutelia.

He told me there had been a debate about whether to arrest Oleg Khinsagov on the spot or let him travel back to Russia for the rest of the promised material. The latter choice would have carried the risk of his disappearing forever, but it also may have led to the recovery of more uranium and possibly shown which mobsters were in his upline. The decision was made to arrest him in the Tbilisi apartment, however, in the hopes that the fish salesman would talk. He didn't. The Georgian government was silent about the incident for nearly a year, but then disclosed the story to the journalist Lawrence Scott Sheets, who has speculated that Georgia was trying to shame Russia into doing more to interdict the loose atomic goods.

Khinsagov had taken his plastic bags of uranium through a mountain border checkpoint called Kazbegi, where the guards had been secretly instructed to turn off the radiation detector and let him through. Yet Khinsagov could not have known this, and something about the state of affairs at the crossing must have led him to believe that he wouldn't get caught. I wanted to go for a look and hired a driver to take me there. This was in late July 2007, a little more than a year and a half after his arrest.

The main road from Tbilisi to the border is known as the Georgian Military Highway, a very old cattle-droving route that was widened and improved in the first decade of the nineteenth century to solidify Russia's hold on its unwilling client state. Maintenance has not been a priority. Potholes and ruts are everywhere, and the road turns to dirt in many places before the broken tapestry of asphalt resumes. Cows find relief from the rain under the concrete roofs of bus shelters, and men wearing brown suit jackets urge horse-drawn carts through the hedged lanes of farm towns. The settlements peter away as the road begins its switchbacking ascent up the first escarpment of the Caucasus Mountains, the high rock wall that traditionally divided Europe from Asia. The air grows sharper up here, and giant valleys open into panoptic vistas of air and stone and grass.

We crested a first summit, passed through the village of Kazbegi, and followed the streamside road into the Darial Gorge, a spot mentioned in Pliny the Elder's *Historia Naturalis* in A.D. 77. This was the site of what he called the Caucasian Gates, a fortified garrison of soldiers that was supposed to protect the cities of the Romans from incursions by Huns and Goths from the northern plains. The gorge was so narrow, Pliny said, that a legion of three hundred soldiers could fend off a much larger army. The gates "divided the world in two parts," he reported, and kept the civilized world safe from the barbarians. Oleg Khinsagov had been through here eighteen months ago with his packets of uranium.

We stopped at the checkpoint on the Russian border, which was temporarily closed while a new American-funded facility was under construction. Five men wearing Georgian military uniforms came out of a nearby barracks. After some translated pleasantries, they took me on a tour of the old facility, a wood-sided shed with cracked linoleum floors. It was marked PASSPORT CONTROL. Out on the roadway, there were two plastic pillars wired up to a gamma-ray detector whose stamp indicated that it had been made by a company in Sweetwater, Texas.

About five hundred cars a day pass through here when the road is open. The detector rings about once a week on average, they said, which means the offending vehicle must undergo a closer inspection. Bananas and ceramics have a way of setting off false alarms. The new checkpoint would have more sophisticated detection equipment, they told me, even though it wouldn't pick up a signal if anyone took the precaution of shielding a block of uranium with a lead sleeve. The soldiers also had a case of handheld radiation detectors, about the size of a telephone pager, that they were supposed to wave over suspicious trucks and cars. They had wanted to show me these detectors, but the case was locked, and they had lost the keys. Above the highway was a giant colorful billboard instructing drivers who felt they were being asked for a bribe to call a special telephone number in Tbilisi to complain.

The guards were friendly and happy to have company. One of them showed me a cell-phone photograph of what he said was the preserved kneecap of Saint George—the national saint—which is kept as a relic in a nearby church. Another pointed to the remains of a castle clinging to a precipice over the stream where a thirteenth-century aristocrat named Tamara had lived. She was said to have been in the habit of seducing lone male travelers and having them beheaded the next morning, their bodies cast in the river. We all went into a small mess hall and shared a lunch of bean soup and thick fresh bread. Somebody poured us glasses of beer. Eventually, a bottle of Chechen grappa came out. They wanted to do shots. We toasted one another's countries.

"Russia is just a big supermarket," one of them said, and laughed. "Whatever you want, you can buy it there. Heroin, uranium, whatever you want."

When I stepped back outside into the cold air, reeling and woozy, I asked the commander, a middle-aged man named Shoto Lomtadze, if he wasn't worried that somebody would simply try to move uranium around this checkpoint, through any one of the other mountain passes that open a hole from Russia into the south. A package of uranium big enough to achieve critical mass would be about the size of a grapefruit and weigh a little bit more than a case of Coca-Cola. The uranium would need to be split in two halves to prevent a premature explosion, but the halves could be easily tucked into a pair of backpacks and simply carried into the country on foot.

Lomtadze smiled and shook his head no.

"It would be much more dangerous for them over there," he said, pointing toward the peaks across the gorge. "They would never make it."

8

RENAISSANCE

The minister of electricity could hardly contain his excitement. He leaped out of his chair and began to pace back and forth on the office carpet in front of me, gesturing to the air while he talked. His impoverished nation of Yemen was destined to have the first nuclear power on the Arabian Peninsula, he told me, and this had to happen within five years. Otherwise there would be a major problem.

"We are in a grave, grave situation," he said. "For us, we have no choice. Coal is too dirty. Dirty, dirty, dirty . . . Hydroelectric? We are a dry country. God did not give me rivers. Natural gas? Nice, if you own it. If you import it, it is five or six cents a kilowatt-hour.

"I have no oil," he said, beating me to the question. "Do you know how much we have? It's declining, and I'm scared we might not have enough for transportation. . . . Right now we are running on diesel and natural gas, paying anywhere from seventeen to twenty-two cents a kilowatt-hour. I am subsidizing electricity here at the rate of twelve cents! The electricity situation is a joke, and it is costing the government a lot of money."

He kept pacing before the cream-colored sofa in his anteroom, building to his peroration.

"There is not a single city in the developing world that is not praying for a huge increase in nuclear power. There is no doubt, my friend, that the nuclear industry is now living in a renaissance."

Yemen's minister of electricity is an exuberant, barrel-chested man named Dr. Mustafa Bahran, who earned the nickname "Dr. Bahranium" in the local press for his evangelism on behalf of bringing a uranium-fueled power station to the edge of the Red Sea as a solution to his country's energy thirst. He has a trim mustache and an intense, probing stare and favors business suits over robes. His doctorate is from the University of Oklahoma, and he proudly refers to himself as a Sooner. He had just returned from a shopping vacation in Dubai when I saw him in his office in August 2007.

Bahran used to be in charge of the grandly named National Atomic Energy Commission, which was responsible mainly for the disposal of old X-ray machines from doctors' offices. After being promoted to the head of the Ministry of Electricity, he created an immediate stir by announcing that his country was "in talks" with at least four nuclear power companies—two from Canada and two from the United States—to host a thousand-megawatt facility. The Yemeni government would have no control over the plant, furnishing just a strip of coastal land and a special unit of the military to guard it. This would make Yemen, as Bahran phrased it to me, "not a nuclear state, but a state with a nuclear plant inside of it."

Such a model has never been tried anywhere, and Yemen seems, at first, like an unlikely place for a pilot. It is a mountain-cleft republic on the southern border of Saudi Arabia, and is often regarded by its neighbors as a throwback—something on the order of the Appalachia of Arabia. Its oil reserves are meager, and poverty is widespread, though it was not always that way. The pastoral tribes who settled here had a monopoly on the trade in spices, especially frankincense and myrrh, up until the second century B.C., and the resulting wealth helped build up the ancient kingdom of Sheba, whose queen is singled out for praise in the Bible and whose economic reach was a source of envy for the Romans. Ptolemy called it Arabia Felix, or "Fortunate Arabia." The people here came under the influence of the Persians and were one of the first foreign entities to embrace Islam, with many converting in the seventh century when the Prophet, Muhammad, was still living. But tribal warfare and power struggles prevented the reestablishment of a unified kingdom, and large portions of the region were run from afar by a series of Egyptian and Ottoman hegemons and then the British Colonial Office until the late twentieth century, when a civil war split the region for more than two decades. It is still said that a Yemeni's true patriotism lies not with the state—generally seen as an artificial construct—but with his extended family in the desert settlements, whose succor and protection meant the difference between life and death in the ancient caravan days; it sometimes still does. If fighting broke out between families, one man in the capital told me, "Half of the government's offices would empty out overnight" as men rushed back to defend their hometowns.

Yemen's rough-edged topography of mesas and hermit valleys seems almost designed for localized warfare, with villages atop the well-protected high ground. Many have public basins chipped out of the mudstone, a kind of natural well to store rainfall for siege. Yet there is also a strong tradition of diplomacy and truces, with complex peacekeeping arrangements, and poems to commemorate them—a particular kind of national opera. Kidnapping is regarded as a fair tool of negotiation, and those who are taken captive are invariably fed lavish meals and treated as if they are honored guests. The federal government keeps a grip on power through a complex series of appeasements and tributes to rural family interests. In this sense, the presidency is like a clan that happens to occupy the capital.

Even in its capital, Yemen seems like a place stuck in antiquity. The core of the lovely city of Sana'a is a district of medieval adobe towers, built in the thick-walled style of Chicago's Monadnock Building amid a twist of cobblestone streets, where eight-hundred-year-old mosques share the same narrow passageways with woodworking shops and the carts of apricot salesmen. The shadows of lamps on stone and the echo of unseen footfalls after dark give the place the aroma of assignation. But a sexual Jim Crow system is in effect, and men who speak with or touch a woman outside their immediate families are asking for trouble. Arranged marriages are the norm: Cousins at some remove usually marry each other in order to keep the clan structure intact. Yemeni women may appear in public only when wearing a black body gown and a *niqab*—a veil that covers the whole face except for a narrow slit for the eyes. Almost every residential window in the Old City is frosted in order to preserve the privacy of women in their houses.

The men usually wear a white sheath garment with a curved ceremonial dagger called a *jambiya* tied around the waist. Though often compared to a phallic symbol, the dagger functions more like a necktie—an ornament meant to signify formality and respect. A big silver coin called a Maria Theresa thaler is still exchanged in some market stalls, though not so much now. First minted in the eighteenth century and bearing the visage of the buxom archduchess of Austria, the coin became the most respected currency in the Arab world. Some merchants here used to accept no other payment for foreign transactions. The coins were legal tender until the early 1960s, which is also when slavery was finally outlawed. (A curiosity: The silver for some of these first thalers circulated in Yemen was mined from St. Joachimsthal, the birthplace of uranium mining.)

Most of the arable land in the countryside and a huge portion of the national water supply is dedicated to the growing of khat, a small leaf that, when chewed in handfuls, provides a sensation similar to that of downing two quick double espressos. Khat is considered a hazardous drug in most other nations, including the United States, where its import is banned. An estimated 75 percent of the men in Yemen (and an unknown, lesser number of women) make chewing khat a daily affair. The leaf is actually sucked rather than chewed, but the "khat chew" has an exalted place in the Yemeni heart as a social ritual among friends and would-be enemies. By 3 p.m. on most days, the right cheeks of most men are bulging as though they have placed a racquetball in their mouths. The government has made it illegal for public employees to use khat on the job, but the rule is widely ignored.

Yemen is an American ally and an official partner in the "war on terror," but the ideology of al-Qaeda has made inroads in some of the rural areas. In October 2000, two young men loaded a bomb into a small skiff and motored out to the edge of the warship USS *Cole* docked in the harbor of Aden. The resulting explosion tore a giant gash in the hull and killed seventeen American sailors. This was followed up by attacks on the few oil installations inside the country. Osama bin Laden, whose ancestors came from Yemen, made a taped speech welcoming the carnage. The government jailed dozens of suspects and was embarrassed when twenty-three of them escaped by digging a tunnel from the basement of their prison to a nearby mosque. Another set of al-Qaeda bombs went off at a shrine to the queen of Sheba in July 2007, killing seven tourists and two local people.

Most Yemenis were horrified by the bombings, and some diplomats privately accused the government of reaching a private understanding with al-Qaeda—no crackdown in exchange for no attacks against the fragile government—and of treating the terror group like any other desert family that must be flattered and appeased. It was in this environment that Mustafa Bahran announced his plans to take the country nuclear with the aid of a Western company.

There are many in Yemen who consider the idea a bad joke and Bahran a delusional egomaniac. But Yemen is not the only Islamic country that has agitated for a nuclear future. Saudi Arabia, Egypt, Algeria, Morocco, and—most spectacularly—Iran all have stated their desire to plug their grids into a uranium reactor, creating a "me too" effect across the region. The president of Yemen, Ali Abdullah Saleh, was reelected in 2006 at least partly on a promise to deliver nuclear energy to light the rural settlements, desalinate more seawater for agriculture, and ease the frequent and embarrassing blackouts in the capital. "This is no longer election propaganda," he said. "This is serious."

He was speaking the language of "nuclear renaissance," the catchphrase used by atomic energy advocates to signal the start of a new dawn of plant building. That even a place such as Yemen has linked its national aspirations to uranium must be regarded as a bellwether for an industry that is ready at long last to shake off sinister images. There is a growing suspicion that the soft old protocols of the toothless Kyoto Treaty will be a matter of history within twenty years, if not ten, and that a forced emergency reduction in carbon emissions will make nuclear energy indispensable because of a simple matter of physics: The fission of uranium emits no greenhouse gases. What was considered the epitome of filth twenty years ago is suddenly looking clean. Several leading environmentalists have reversed their longtime stance against the technology and come out in favor of the thing they once reviled. Such a selling point has become a central message for advocates of nuclear power, many of whom have called for a doubling of the international fleet of nuclear reactors in the name of the oncoming fight against global warming. China has gone even further. With a skyrocketing population and some of the dirtiest coal-burning plants in the world, it has plans to quadruple its own fleet to handle the crushing demand for power. Its state mining companies are combing the Sahara for uranium reserves.

This new confidence in a technology considered dead and buried raises the same questions that were pertinent in the 1950s during the Atoms for Peace campaign that spread reactors around the world. How do you dispose of the waste? How do you keep the plant from being a military target? How do you make sure the plutonium by-product isn't being secretly taken elsewhere and packed into the core of a bomb? Can global security really be guaranteed with so much uranium stacking up in the developing world?

Security has a localized definition, too, and in Yemen that translates as the survival of a politically moderate government that cannot deliver the goods to its people. Nowhere is that on starker display than in Yemen. The intractable poverty and the amateurish (though effective) attacks from rural bands of al-Qaeda sympathizers recently prompted the journal *Foreign Policy* to name Yemen the twenty-fourth most likely country in the world to be torn apart by "violent internal conflict" and "societal deterioration." Most diplomatic observers are convinced that Yemen will be a failed state within fifteen years.

Oddly enough, this is a major rationale behind Bahran's nuclear gambit. It is one more truce in a nation with a long history of making truces. President Saleh has been in high office since 1978 and has reversed earlier pledges to step down. The strategy for easing the unhappiness in the villages involves bringing more power—electrical, not actual—to the people, thus creating the perception that the entrenched leadership has done something tangible. Bread and circuses now come in the shape of a cooling tower. "If I don't have electricity to make energy," says Mustafa Bahran, "I will not survive."

He likes to tell people that if all of the employees at the Yemen Ministry of Electricity were assigned a stationary bicycle with wheels connected to a turbine, they would generate more power in a day than the nation now produces in a week. The current output is now about 770 megawatts, which would be considered adequate coverage for a city the size of El Paso, and Yemen has twenty-eight times the number of people as El Paso. Lights and television are still considered exotic in large parts of the desert interior, and the population there has an average family size of eight. Nuclear power is the only possible catch-up scenario, says Bahran, who assures the skeptical there is no reason to be worried about hazardous waste or security from terrorists.

"I am making a revolution in electricity," he tells me, pacing. And, "I am very, very particular about safety. The waste will not stay in Yemen. The company will take the waste. There will be no reason for panic." The reactor would be owned and operated by a Western company, which would guarantee the security of the uranium fuel assemblies and make sure that all the material was accounted for. The Yemeni military would be responsible for safeguarding the arrival of the fuel rods once they passed into territorial waters, as well as the departure of the spent rods and the plutonium. Even an armed attack on the facility would be harmless, Bahran told me. "You will have a substantial component of the army surrounding that plant. Not even a bee can go through the perimeter. Suppose the fundamentalists take over this country. They would find very little spent fuel in the reactor. They will see only one batch."

Not everyone in Yemen shares his buoyancy.

"This is growth without means," said Abdullah al-Faqih, a professor of political science at the University of Sana'a. "You need a lot of money to do this, and there are a lot of risks. Even talking about nuclear power is not good for us. If we insist on this, we will lose foreign support. It's giving the world a signal, and it will send the opposite message—it will actually drive away investment." Another observer in Sana'a was blunt: "This is a country that cannot even clean up the plastic bags on the side of the road. How can we expect to handle nuclear waste?"

I sat for lunch one day with a man named Ali Nessar Shoueb in his home, which is above a ground-level floor where he keeps a herd of fifteen donkeys. He studies chemistry at the University of Sana'a, but was home on a summer break. We ate flat bread, rice, and chicken while reclining on green floor cushions. In the next room, Arabic cartoons were playing on a television mounted high on the wall. Behind the TV was a Beretta pistol, hanging by the trigger guard from a nail pounded into the mud wall.

Over tea, Shoueb told me the recent al-Qaeda bombings had been disastrous for him, as they had driven tourists away from his town of Thula, a picturesque medieval trading village at the base of a sandstone pillar.

I asked what he thought of the plan to bring nuclear power into Yemen. He told me it would certainly help ease the task of hauling in the weekly supply of water his family and his livestock required—about 530 gallons, by tanker truck. Nuclear energy could help desalinate the water from the Red Sea and make everything cheaper for him. "We need more light and more water," he said. "I think it's a good idea."

He paused. "But I would want it in the village in back of me. Actually, ten villages away."

There was a bust of Homer Simpson in the lobby of the headquarters of the World Nuclear Association in London.

The joke may have been lost on some visitors, but for those who watch *The Simpsons*, it was a clever bit of self-effacement. Homer is the oafish patriarch of the cartoon family, and his profession is a standing gag: He works in the control room of the local nuclear power plant.

"Can you imagine a bigger idiot?" one staff member asked me. "He's the public face of the industry, so we thought we'd put him out front."

This kind of swagger is now possible for the first time in decades. China and India have developed an enormous thirst for electricity, and the demand for global kilowatt-hours is expected to double in the next three decades. Plans for nearly two hundred new reactors are under way in nations all around the globe—including more than a dozen in the United States, where construction has been static since the early 1980s. A new generation of reactor developed in Germany—the Pebble Bed Modular Reactor—uses fuel pellets the size of tennis balls and can operate at double the temperature of older models. It is the design currently favored by China, which has plans to build thirty of them. This design is also being promoted as the perfect "entry-level" reactor for beginner nuclear states such as Yemen. At the end of 2006, President George W. Bush negotiated an agreement to sell large quantities of enriched uranium to India. Critics called this a flagrant violation of the Nonproliferation Treaty, which the weaponized state of India has never signed, but Bush emphasized instead the need for the world to develop "clean and safe energy." All of this has spread euphoria inside the nuclear trade, and a renewed sense of mission. Up until recently, petroleum was where all the action was in the energy business. The nuclear guys were considered the dullards, parishioners of a dying church. No more.

"Uranium can quite literally save the world," said John Ritch, the director of the World Nuclear Association. "This is a remarkable mineral, and humanity has found reliable ways to turn it to the betterment of everyone. It is surrounded with myth and fear, but it is also surrounded by constructiveness."

Ritch told me his ideal outcome would be 8,000 nuclear reactors operating within this century, up from the current worldwide level of 440. "We have only begun to tap the world's uranium reserves, and the use of uranium generates a minuscule amount of waste that, with scientific assurance, can be dealt with safely," he told me a later e-mail. "There is no doubt that bad actors can abuse nuclear technology. But those bad actors will be there in any case, and we must target our efforts on thwarting them, whether in North Korea or elsewhere. Meanwhile, we have built a high wall between the peaceful and illicit uses of nuclear power. Today, we can expand the use of nuclear power twenty-fold without increasing nuclear danger a bit," he added.

His office looks out onto St. James's Square, a small park ringed by some of the most blue-chip properties in London. The address signifies peerage, clout, and prelapsarian money. The Queen Mother spent her girlhood years in the Georgian town house across the street. At the far end is Chatham House, now a foreign policy institute, but once the home of prime ministers Chatham and Pitt. The world headquarters of Rio Tinto—the stalwart of the old price-fixing Uranium Club—is at No. 6. Down the block at No. 1 are the executive offices of British Petroleum. On the east end is the town house occupied by Supreme Allied Commander Dwight D. Eisenhower when he was doing the logistical planning for the D-day invasion in the spring of 1944, twelve years before he would initiate the Atoms for Peace program. The World Nuclear Association itself occupies the most modernist building on the square, a rectangle of metal and glass. Once called the Uranium Institute, the organization changed its name several years ago to the more universal-sounding World Nuclear Association.

Even this name isn't perfect, said Steve Kidd, the genial director of strategy and research. He believes it has unsavory connotations.

"If you say 'World Nuclear,' the next thing you would think of is a bomb," he told me. "Nuclear power was a mislabeling. It should have been called 'fission power.'" The pronuclear author William Tucker has argued for another moniker: "terrestrial energy," because uranium is, after all, a product of the earth.

This echoes a frequent lament among those in the business—that of a persistent image problem around the word *nuclear*. The term literally means the "manipulation of the nucleus." But history has given it a range of unflattering images: glowing green stuff, genetic mutations, Hiroshima, global warfare, meltdowns, death. Perhaps the signal event was the near-catastrophe in 1979 at the Three Mile Island plant near the capital of Pennsylvania. A series of mistakes by poorly trained technicians (at one point, a blinking light on a control panel was covered up by a yellow maintenance tag) caused half of the uranium fuel to melt and a large bubble to form inside the reactor shell. There was no such shield at Chernobyl in the Soviet Union, the scene of a much worse disaster in 1986. A botched test of a turbine generator caused a steam explosion and a melting of the fuel rods. The graphite started to burn, and fifty-six firefighters who rushed into the atomic volcano paid with their lives. A cloud of radiation drifted as far as Norway.

The extent of the fatalities from such incidents is still being argued about today (the pronuclear side counts only the Chernobyl firefighters;

its opponents estimate thousands more from thyroid cancers), but the net effect was to galvanize the environmental objections to nuclear power and make the regulatory and approval process even more lengthy for putting new reactors online in the United States. The average wait time is still about seven years. This lag is a matter of some frustration at the World Nuclear Association.

"I don't have a lot of regard for environmentalists," Kidd told me. "They pounced on nuclear power thirty years ago as an easy target. Now it turns out they should have focused on the automobile. They don't like big cities or big organizations, or working long hours. They want to ban cheap flights for my holiday. They want to take us back to the Stone Age. They are silly people who want to stop development."

One of the manifest ironies of the "nuclear renaissance," though, is that it relies on an image of atomic power as a green technology—a clean alternative to the coal-burning plants that have long been the world's electrical mainstay. Coal is a particularly dirty and dangerous fuel in China, where an estimated five thousand miners die in accidents every year. That nation is now pouring up to about 26 million tons of sulfur dioxide and 3.2 billion tons of carbon dioxide into the atmosphere each year, creating pollution so thick that in the worst areas people must drive with their lights on during the daytime. Yet China must also feed an overdrive economy, expanding 10 percent each year. An aggressive nuclear strategy has been the obvious answer. Unlike harnessing the wind or the sun, uranium power is here right now and ready to go. And a single ton of raw uranium provides the same electricity as twenty thousand tons of black coal.

"One of the fundamental imperatives in the world is to harness the source of this cheap energy," Kidd told me. "I have no doubt we'll have three thousand reactors, and I've heard projections of ten thousand."

The American energy policy crafted in secret during Bush's first term was generous to nuclear power, allocating up to $13 billion in subsidies and tax credits to the industry, with the aim of starting a burst of reactor construction within the decade. The green argument has swayed some historical opponents, among them U.S. House Speaker Nancy Pelosi, who told a congressional committee, "I have a different view of nuclear than I did twenty years ago. I think it has to be on the table." The editorial page of the *New York Times*, once skeptical, said: "There is good reason to give nuclear power a fresh look. It can diversify our sources of energy with a fuel—uranium—that is both abundant and inexpensive." A cofounder of Greenpeace, Patrick Moore, used to deliver rants against what he called "nuclear holocaust," but he has now come out in support and is a paid consultant to the industry.

The most surprising defector, however, is James Lovelock, who is most famous for his "Gaia hypothesis," which says that Earth is a living organism that breathes. He has since joined a lobbying group with a name that would have seemed out of a *Saturday Night Live* skit of twenty years ago—Environmentalists for Nuclear Energy—and has come out loaded for bear. "Opposition to nuclear energy is based on irrational fear fed by Hollywood-style fiction, the Green lobbies, and the media," he wrote in the London *Independent*.

All of this has been excellent ammunition for the new promoters of uranium, who find their clearest expressions, once more, within the vocabulary of apocalypse.

"The fact of this planetary crisis should no longer be a matter of psychological or political denial," John Ritch said in a 2006 speech. "For our best Earth-system scientists now warn, with ever increasing certainty, that greenhouse gas emissions, if continued at the present massive scale, will yield consequences that are—quite literally—apocalyptic: increasingly radical temperature changes, a worldwide upsurge in violent weather events, widespread drought, flooding, wildfires, famine, species extinction, rising sea levels, mass migration, and epidemic disease that will leave no country untouched."

The fresh excitement about nuclear power sent the price of uranium rocketing upward. In the spring of 2007, utilities found themselves paying up to $132 a pound for spot deliveries, a price that would have seemed like fantasy just a decade ago, when uranium cost a tenth of that. Uranium became a viable commodity for the first time since the 1920s radium boom at St. Joachimsthal, as several New York hedge funds acquired quantities of yellowcake and kept them in storage, betting that the price would go up even further. The run-up was aggravated by heavy rains at the Ranger mine that shut down production for a week and heightened the impression of a gap between a stagnant supply and voracious global demand.

Compounding the problem was the lack of talent in the uranium business, which had effectively gone to sleep. No sane graduate student of geology would have picked uranium as a specialty after 1985: It would have been like learning Morse code instead of Linux. The academic sciences made a decisive turn away from nuclear after Three Mile Island and the disarmament battles of the early Ronald Reagan years.

"I went to Cornell University to learn nuclear engineering," said Joe McCourt, who is today the head of a uranium brokerage. "I wanted to make the world a better place, provide cheap energy for the masses, and that kind of thing. And then we just got vilified. Cornell ended the program and dismantled their reactor. And so now there's a big generation gap in this field. We lost a generation."

This meant that a lot of old uranium hands from the 1970s suddenly started getting phone calls at their retirement haciendas in Florida when the price started to rise. A number of them were persuaded to get back into the game. One of them was Bill McKnight, now seventy-one, who went back to work for Uranium Resources, a company he helped cofound more than three decades ago. Though he now works as the vice president for exploration, he favors a driller's jumpsuit to a coat and tie. McKnight got his start hunting uranium on south Texas ranches in the 1960s and once processed yellowcake in a horse trough, using a kitchen-faucet water softener for an ion exchanger, until his employer at the time, Mobil Oil, told him to quit it. The utopian promise of uranium still excites him much more than petroleum.

"I want to do something that contributes to society," he told me. "I served in the military for the same reason."

His wizened visage did not make him stand out particularly at the 2007 Global Uranium Symposium at the Omni Hotel in Corpus Christi, Texas (this year's theme: "Taking U into the Future"), where nearly every participant was older than fifty and the lunchtime speeches tended to open with jokes about bum knees and bald heads. But the energy inside the convention hall was palpable. "This is like a bunch of kids getting together to build a tree house," one geologist told me. Anybody who knew anything about uranium was getting a nice salary and lots of attention.

The president of McKnight's company, Dave Clark, cautioned everyone in the room about feeling overconfident. There was a time, he said, when people on Wall Street didn't know how to spell uranium. Now that the times were good, it was well to recall the old prayer of an oilman: Please, please let there be another price spike and I promise this time I won't fritter it away.

"All these guys are going to be coming in when the money's good, and they'll be gone when the money's bad," Clark said. "The market is going to obey the laws of gravity, and there's going to be a change in perception. There is no shortage of uranium. No reactor is going to shut down for lack of it."

He was restating a key precept in the energy trade: Nuclear plants are expensive to build, but cheap to fuel (it works the other way around with a coal facility). The cost of uranium is barely an afterthought for most utilities. And if the nuclear renaissance can become more fact than hype, an even more permanent thirst for the "bad-luck rock" will have been fixed in place.

"I'm sorry, but the genie's out of the bottle," said David Miller, the president of Strathmore Minerals. "I don't see why we can't use it to bring the poorer countries of the world up to a better standard for all mankind. You're not going to do that with coal or solar. Nuclear power is a savior of the world."

He then repeated a favorite maxim of the uranium business, one in play in the American West ever since the days of Charlie Steen and the fat government bonuses: "Coal was the fuel of the nineteenth century, oil was the fuel of the twentieth century, and nuclear will be the fuel of the twenty-first century."

Much of this century's uranium will eventually pass through the far southeastern corner of New Mexico. This is a state with abundant uranium reserves in the mountains near the town of Grants, two national laboratories, and a history with radioactivity that goes back to

Los Alamos. More important, it is the home state of U.S. senator Pete Domenici, the former chairman of the Energy Committee and the self-described "chief nuclear apostle" of Congress, who lobbied heavily for the nuclear subsidies in Bush's energy policy. He helped lure a consortium of some of the biggest utilities in the country into building a huge $1.5 billion enrichment plant near the petroleum town of Eunice, where tall yellow signs at all four highway entrances proclaim FRIENDLY PEOPLE, PROUD TOWN.

Eunice is on a reddish plain overlying the northern shores of an underground sea of crude oil discovered in the 1920s and responsible for the thicket of pump jacks, tanks, and electric wires that cross the flat scrub, along with the yucca and the prairie grass. The rest of New Mexico refers to this region as Little Texas, and not generally with fondness. The air is scented with hydrogen sulfide, a by-product of the natural gas emissions that blow through town when the breeze is up. A quail pasture to the east of town will be the site of the new uranium enrichment plant, which had—not so long ago—been on the verge of becoming a nonstarter.

A front company known as Louisiana Energy Services, whose investors included Exelon, Duke Power, and Louisiana Power & Light, and also the European atomic giant Urenco, first had wanted to put this plant near the town of Homer in northern Louisiana, where the residents are mostly African American and the economy is moribund. High-wage jobs and tax revenues were promised. But environmental lawyers got involved and started filing suits. An official who prepared the site-selection study in the early 1990s later admitted under oath that he picked Homer because the houses appeared poor and dilapidated. After an eight-year court fight, during which charges of "environmental racism" were thrown around, the Sierra Club succeeded in getting the building permits revoked.

The consortium next tried Hartsville, Tennessee, where the project was again scotched in the face of local objection, bolstered by the intervention of former vice president Al Gore, who has not embraced nuclear power despite his personal crusade to stop climate change. "I can say with no hesitation that this facility is not in the best interest of Middle Tennessee," he said in a statement. "The accumulation of

hazardous waste may become a never-ending problem for local citizens."
That was strike two.

After eight years of frustration, the consortium finally secured
permits to build near Eunice after Senator Domenici got involved. A
few prominent citizens of Eunice (including a hairstylist, the county
emergency coordinator, and the manager of the Pay-N-Save) were taken
on an expenses-paid trip to Urenco's plant in the Netherlands to see for
themselves how the plant was nestled among the grain farms. One citizen was invited to speak to the Eunice Rotary Club and told stories about
the Dutch people's fondness for bicycles. There was no significant local
opposition, even though the company has not released a detailed plan
for permanent disposal of canisters full of depleted uranium gas. One of
the proposed dump sites is located in a dusty field just over the border
in Andrews County, Texas, where a company called Waste Control Spe-
cialists has been licensed to bury the historic reserves of spent ore from
Shinkolobwe that had been used to make atomic bombs at a plant in Ohio
(the Congo waste was still so radioactive it earned its own code name—
K-65—and was noted for its singular strength).

"I am delighted and proud that the renaissance is in New Mexico,"
Domenici said at the ground-breaking ceremony. He related how he had
told the company "to stop putting up with all this guff and apply to build
the facility in New Mexico." The plant is located just barely inside the
state; its east fence is less than a mile from the Texas border.

I was driven up to the plant's security gate by a company spokes-
person named Brenda Brooks. The poured-concrete barn shell of the Sep-
arations Building Module was directly in front of us, rising up from the
desert floor as if a new mesa had pushed up overnight. This building will
house the centrifuges, the same essential kind of enrichment technology
that A. Q. Khan had stolen from the Dutch and sold to Iran, Libya, and
North Korea.

Everybody with access to the inner parts of the centrifuges at Eunice
will have to acquire a top-secret "Q" clearance, said Brooks, and all of
them will be working for a special subsidiary called Enrichment Technology of the United States and have limited contact with the rest of the
construction team.

"A lot of countries would like to get their hands on this," she told me. "There's a huge responsibility with having this technology, and we take that very seriously."

There are only six other plants like this in the world today, and only one now operating on the North American continent. It has more business than it can handle.

To enter the tubular maze of the United States Enrichment Company outside Paducah, Kentucky, you must first submit to a criminal background check and be walked through a huge scanning machine at the front gate. Encircled by cornfields and down the road from a failed bedroom community called Future City, this plant houses what might be called the holy grail of uranium: the top-secret process that turns worthless powder into fuel pellets. I came here on a summer afternoon and was taken on a limited tour by the head of public relations, Georgann Lookofsky.

The plant had been built in semisecrecy on top of an old munitions factory in 1952, just as the mining fever was gearing up in Utah and Arizona. All of that uraniferous soil needed a place to be processed into weapons, and the original plant at Oak Ridge was overtaxed. The impoverished Ohio River Valley was eager for jobs, there was a lot of water, and it was far from the seditious influence of big cities. Most of America's uranium entered this facility in railcars and left it ready to be fashioned into the shells of nuclear weapons.

The uranium was enriched through the same method pioneered at Oak Ridge, a method now considered hopelessly antiquated. Uranium hexafluoride gas is forced through a series of chambers webbed inside with wire-mesh screens that filter out the infinitesimally lighter atoms of U-235. (The gas is corrosive and is hard on industrial pipe material. Much trial and error had gone into making the screens, and it has never been disclosed exactly what they are made of. This was one of the closely held secrets of the atomic era and is still classified.) The entire facility is powered by a coal-burning plant in the nearby town of Joppa, Illinois, directly across the Ohio River, that provides a thousand megawatts, more than the entire electrical output of the nation of Yemen.

I was allowed to walk through the stadium-size room underneath the cascades where the exhale of the compressors drowned out every other sound. Each compressor consumes about as much power as a freight locomotive as it spins a ringed cylinder inside the conversion chambers on the second floor. The cylinder acts a bit like the blades of a blender, forcing the uranium gas against the screens to divorce its fissile component from the surrounding lifeless isotopes. There were lines painted on the floor showing where visitors were allowed to step. A sign said DO NOT CROSS BLUE LINE. Oozing through the braiding of pipes above my head was the concentrate sought by ambitious nations, the most destructive fruit the earth could yield. This fuel leaves Paducah packed in kegs about the size of a hot-water heater, each of them weighing about two and a half tons.

There have been terrible mistakes made here. Uranium waste was burned out of the smokestacks at night—a substance called "midnight negatives"—and radioactivity has been found in the soil nearly a mile away. The Energy Department concluded that nearly sixteen hundred tons of atomic weapons parts, some of them contaminated with enriched uranium, had been scattered around the plant in various locations. The workers here were never told about the dangerous conditions until a 2000 investigation by Joby Warrick of the *Washington Post*.

The cleanup at Paducah will take an estimated seventy-five years, but the plant will have been shuttered long before that. The United States Enrichment Company is building a $1.7 billion replacement facility up-river in the town of Piketon, Ohio, and this one will run on the more modern method of centrifuges, the process favored by Pakistan and Iran. President John K. Welch assured his shareholders the plant will be the most efficient ever built.

"A renaissance is underway in the nuclear power industry, and the signs are everywhere," he said.

Old schemes were being resurrected in obscure parts of the world. Geological maps that had been moldering away in ministry filing cabinets from Azerbaijan to Zambia were being pulled out and given a fresh look.

A lasting truism of the uranium trade is that the best place to find a new mine is next door to an old mine. Dead uranium zones were breathing again, and one of them was in the desert country of the American Southwest, where Charlie Steen had made and lost his fortune half a century before.

During the last boom days of the 1970s, a company called Energy Fuels Nuclear had run eight shafts in a high plateau of cliffs and sagebrush called the Arizona Strip, which lies several miles north of Grand Canyon National Park. Its claims went fallow after the market crashed and the company went bust. Now the remains were being eagerly picked over by a new wave of geologists and prospectors.

I went out to the strip to meet a friend, Walt Lombardo, who has a dark mop of hair and wire glasses and used to head a regional office of the Nevada Division of Minerals. He and his wife, Sandy, now own a bookstore specializing in earth science topics on the outskirts of Las Vegas, and I befriended them after they hosted a reading for me a few years ago. They have been closing their bookstore on the weekends to grab some of the more promising uranium spots still left for the taking. Walt took me walking down a streambed where rainwater had drained from the previous day's storm; the water had painted a damp red streak on the sand.

"Look there, you can see how those beds are plunging," he said, pointing to the edge of a canyon wall showing a banner of sandstone layers trending downward. "You want to see those all over the area. We did some initial reconnaissance here and it looked good."

The object of his hunt is a depression known as a breccia pipe, shown to be the occasional host of radioactivity. It is basically a geological trash hole—a hollow tube shaped like a carrot into which a hodgepodge of rocks and debris has tumbled during the last ten million years. The origin of these tubes is a mystery, but the most accepted theory is that they were carved by hot water erupting upward and then became clogged with limestone and other silica, which had eroded, leaving a cavity that caught rocks and other debris. Groundwater containing a soup of liquefied metals had flowed through a few of these old channels and left residue, which may or may not have included uranium.

Finding a breccia pipe requires thinking in several dimensions. You must first have some photographs of the area taken from an airplane. You then must study them for any signs of an oval depression in the ground, where soil might have settled a few feet into the mouth of one of the hidden tubes. Then you must take your jeep or truck out to that spot and look hard for trending that suggests gravitational pull, as though a hand from hell has yanked down on the earth's surface—as if yanking a tablecloth through a knothole.

"You don't look at a particular outcrop: you have to look at them all," said Walt. "You have to see these beds dipping toward a common center." One hopeful landscape is a small plain with buffalo grass but no sagebrush. The root system of sagebrush generally does not do well on top of a breccia pipe. Another trick: Always stake windmills. A rancher surely stuck it there because there was good water underneath, and water collects in breccia pipes.

The only way to tell a breccia pipe for certain is to pay a contractor up to $15,000 per day to drag an apparatus called a diamond drill out to the site. The drill looks a little like an oil derrick, and the physics are just the same: An ugly metal bit with edges made of industrial diamonds (the best cutting material in the world) is screwed onto the end of a hollow steel rod and rammed thirty-three feet into the sandstone. Then it must be hauled up, another rod screwed on, and then rammed down farther; more rods are added as the bit chews down to where the uranium might be layered. This yields a slim tube of rock called a core sample, which is about the circumference of an apricot and can be crushed and analyzed. Radioactivity is a prospector's friend, as uranium announces itself louder than any other mineral. A small machine called a spectrometer can be lowered down into the drill hole to see what's glowing.

But drilling into a breccia pipe is a gamble for any mining company, as only 1 in 8 pipes has captured any uranium at all, and only 1 in 150 bears it in rich enough quantities to justify the expense of digging for it. The debris inside the pipe is tough on the rods; breakages are routine. Roulette has much better odds. The Arizona Strip is nevertheless the scene of a small-scale uranium rush. A resurrected mine called Arizona One is already back in production, hammering rock that had last been touched twenty years ago. Claims to the local office of the Bureau of Land

Management quadrupled in 2007, under company names such as Liberty Star, Lucky Irish, and U.S. Energy.

"These are small guys making the discoveries," Walt told me. "The big companies have preconceived notions of what they're looking for and they have no creativity. There is no room for the dreamer or the artist."

Energy Fuels Nuclear had been out here first, armed with a team of cowboy geologists and a juicy contract to supply a reactor in Switzerland. The proprietor was a plug-shape restaurant owner from Rawlins, Wyoming, named Bob Adams, who had caught the fever after reading a newspaper story about Charlie Steen. Adams knew how to fly—he had been a bomber pilot in World War II—so he began making exploratory flights around the Wyoming outback. Before long, he found a radioactive anomaly on a dry plain some distance north of Rawlins, not far from the wagon ruts of the Oregon Trail, where wagon trains had passed a century ago. He secured the bankrolling for a mill from a group of Colorado investors, and soon a cluster of prefabricated trailers sprang up—named Jeffrey City for one of Adams's early investors. The *Denver Post* called it an "atomic age frontier town." An Army-style barracks was hastily erected for the bachelor roughnecks up from Texas and asphalt streets were thrown down on the hardpan; they bore names like Jackpot, Uranium, and Jackalope. Townhouses with hideous orange carpet were added later. Life outside the mines reportedly grew so boring and drug-addled that the miners started swapping wives just to stay entertained. By 1972, Adams had become rich enough to buy a coal company, and he started poking around for more uranium on the plateau of cliffs and sage plains called the Arizona Strip.

The strip was one of the last sections of the American West to be settled, and it was then—as it is today—lonely country. Mormon pioneers had scouted the area in the 1850s and found it unpromising, though they did haul timber from the slopes of Mount Trumbull for the construction of their first temple. One of the only towns to take permanent root here was a place called Millennial City, inhabited almost exclusively by a band of polygamist families. They had left Salt Lake City in the 1930s to escape the harassment of the mainstream Mormon church, which considered them apostates. They came to the desert to prepare for the final judgment, for which only they were worthy, because they had defied the Utah constitution and kept God's revelation that a man should have multiple wives. "Hell, if I had to live out here I'd want more than one wife myself," said Arizona's first governor, George W. P. Hunt, after an official visit. Yet the settlement at the base of the Vermilion Cliffs was advantageous: It straddled the Utah-Arizona border, which meant that quick escapes could be made from sheriffs of one state or another. When apocalypse failed to arrive, the town changed its name to Short Creek and settled in for the long haul. The place is known today as Colorado City and is the home of the ten-thousand-member Fundamentalist Church of Jesus Christ of Latter-day Saints.

Bob Adams found the young men there to be excellent ore muckers and hired several of them to work in the breccia pipes. After they took all of their wives to the Christmas party and nearly broke the entertainment budget, a new policy had to be crafted: All of a man's wives were welcome at company picnics, but only one could be taken to dinner with a client. And only one could go on the insurance policy as a dependent. Energy Fuels Nuclear grew powerful enough in the late 1970s to acquire all the competing claims on the Arizona Strip. It serviced its eight main holdings with a fleet of three helicopters, based out of the small town of Fredonia, Arizona. The polygamists tended to avoid liquor, but other employees did not.

"People played hard and lived hard," remembered sixty-three-year-old Roger Smith, the manager at the old Pigeon Mine. "It was a jumping, jumping time. People came to work with guns on their hips. In those days you could run a mine inspector off the property. Today, you'd go to jail for that."

Smith wears Elvis-style sideburns and black Reebok athletic shoes. A buck knife rides on his belt. I met him in sunbaked Fredonia, in the yard of his decorative stone company, where he was driving a forklift. He took me inside the office and offered me a beer. Tacked above his air-conditioning unit is a bumper sticker: EARTH FIRST! WE'LL MINE THE OTHER PLANETS LATER. On another of his office walls is a photograph of one of the old uranium mines on the strip: a cathedral-size hole blasted into the sandstone, with dirt ramps for the wagon loaders.

Smith had a reputation as a tightwad, to the occasional resentment of his employees. One of them absentmindedly left his jackhammer in the bucket of a loader. It accidentally got sent to the mill with a load of ore that evening, and the hammer was turned into a metal pretzel: "Smashed to shit," recalled Smith. But the serial number was still legible, and Smith figured out who it had belonged to. He deducted its cost from the paycheck of the forgetful man, who took revenge by dumping loads of waste rock into the ore pile to dilute its value.

These pranks did not affect the bottom line for Bob Adams, who had become an extremely wealthy man before he died unexpectedly in 1982. He had just finished a room-service dinner in the Essex House hotel in New York City and was settling in to watch television when he had a massive heart attack. The medical examiner's report indicated he was alone in the room.

He had owned a ranch in the ski town of Steamboat Springs, Colorado, and the *Steamboat Pilot* memorialized him on the front page, remembering him as "a giant of a man" and "a citizen concerned with his world." The paper went on to note that "he died, as he had lived, with flare [*sic*]" in his luxury hotel. Adams's death came at the beginning of the long slide in uranium prices that would eventually kill the mines on the Arizona Strip, and with them, his company.

Jeffrey City quickly turned into one of the eeriest ghost towns in Wyoming. Houses were abandoned overnight with the disco-era furniture still inside, which rotted slowly in the desert heat. Empty concrete pits, once the foundations of homes, lay open to the sky. I paid a visit there on a chilly spring night and wandered down the dark hallways of the bachelor barracks, which is full of rotting wood and broken linoleum. A sitting room had been partially consumed by fire. Piled in one corner was a tattered *Sports Illustrated* from 1977, a catalog for eight-track tape decks, and a religious tract titled *Love's Dimensions*.

Bob Adams's old manager, Roger Smith, now works as a consultant to some of the new uranium outfits prowling the Arizona Strip, but he has not formally joined any of them. He is known in Fredonia as one of the eminences of the last wave, as well as a gregarious cutup, and his opinions on the nuclear comeback are taken seriously.

"It's a happening thing," he told me. "You are not going to see wind or solar anytime soon. We have to start thinking about what's clean and safe and inexpensive. Everybody likes to kick back and have a cold beer in an air-conditioned house. And where does that come from?"

Most of the mineralized breccia pipes are still unexplored, and large sums of capital are being wagered in the hopes of finding more. The ore grades on the strip historically averaged 0.2 percent—lower than Moab's—but still rich enough to be worth digging during price spikes. And public land is free for the taking, or nearly so. The General Mining Act of 1872, a law that hasn't been substantially revised since its passage under the presidency of Ulysses S. Grant, says that a prospector must affix wooden posts at the corners of an area no bigger than six hundred feet by fifteen hundred feet. You pound in a central stake known as a discovery monument and at the base leave a waterproof container holding a document with your name and address and the surveyors' coordinates of the land, all written underneath a magisterial opening: *Notice is hereby given*. . . . A duplicate copy must be mailed to the county recorder. And that's all. You now own everything under the surface. And if you're lucky, a mining company will decide there's enough uranium hiding in there to be worth a deal.

Walt and Sandy were busy hammering posts into a sloping patch of caliche when Walter spied something interesting on the ground. He carried it over to me. It was a flake of stone shaped like a teardrop, colored ivory, and thin and sharp on the edges. It looked as though it had been worked with another stone. This was chalcedony, a crystalline form of silica and a favorite source of tool points for the local Paiute Indians and, before them, the Anasazi, who vanished in the thirteenth century. The stone may have once been tied onto an arrow shaft and fired into the ribs of an unlucky antelope, whose bones had long since gone to dust. Or it might have been a reject. There was no way to tell.

"We find these out here from time to time," he said. "This land has been inhabited for the last five thousand years."

A few tickles of radioactivity can be turned into cash, and the best place to do it is the Canadian city of Vancouver, the financial capital of the new uranium rush and a historic tank of sharks.

At least four hundred uranium companies, known as juniors, were operating in Canada in the autumn of 2007, and most were squirreled inside various rented offices in downtown Vancouver, cloaked behind a forest of nameplates on an oak door and a shared secretary to turn away visitors. They have Internet sites advertising possible future drilling in such far-flung spots as Argentina, Peru, and Kazakhstan. Others aimed for ground in more established uranium fields such as Wyoming, Niger, or the heavily staked Athabascan Basin in Saskatchewan. There are often scenic pictures of wilderness decorating the home pages of the Web sites, as if time-share condominiums were being sold instead of stock certificates.

The quality of *dreaming* is the same, though, because almost none of the juniors will ever see an ounce of uranium. The object is to convince a brokerage house that a ghost of a chance exists of there being some ore lying underneath a claim, enough so that one of the major players, such as Rio Tinto, Cameco, or International Uranium Company, might be sufficiently paranoid to buy it for fear of the property's going to a competitor. And everyone involved (except, perhaps, the investor) knows full well that only about one in every ten thousand claims will ever come near to being a mine.

Making this pitch requires what brokers call a "story"—a quick verbal summary of where the claim is located, what its historic reserves might be, what kind of drilling has taken place so far, and some brief biographies of the management team. The story is what moves the stock, not the actual uranium that might be pulled out of the ground, which is a faraway concept almost irrelevant to the entire process. The old saying in Hollywood goes that anybody is a "producer" who has a quarter to put into a pay phone, and the bar is set low in a similar way in Vancouver, where anybody who can hire a geologist to sign a disclosure and a promoter to hype the stock (euphemistically called V.P., investor relations) can be the president and CEO of a uranium company.

"Most of these 'stories' are fantasy," said Jim Cambon, a mining executive. "And they aren't *lying* so much as believing their own bullshit."

If the brokerage house can be persuaded to write an analysis and offer shares to its clients, the market will notice and the stock price will rise a few cents, leaving the principals in an excellent position to sell their own shares at peak. They might retain an interest in case of a second spike, but the point is still to make a quick payday and move on to the next thing. The brokerage makes its money from fees and is rarely in a position to object to a dubious offering. No buyout occurs, no revenue is produced, and no mine ever appears. This is known as mining the markets. The more egregious cases are called pump and dump. It is sometimes illegal, occasionally investigated, and almost never prosecuted.

"It's hard to differentiate between the smoke and mirrors and the legitimate exploration," said Robert Holland, the chief geologist for the British Columbia Securities Commission. "We've seen cases where companies incorporate and get a rinky-dink piece of property, do a ten-thousand-dollar work program, and the disclosure is all there. They're able to get listed."

Vancouver is a "city of glass," in the words of the novelist Douglas Coupland, a rainy peninsula of Finnish-modern apartment towers and espresso bars and seaplanes owned by Hong Kong billionaires. It had become a haven for shady stock deals partly because of lax securities regulation, but mostly because of its physical setting at the mouth of the Fraser River and at the foot of the Coast Mountains. This was the gateway to the furs and lumber and gold inside the high mountain valleys of British Columbia in the first wave of natural resource exploitation in the 1860s. Sawmills and freight yards sprang up along False Creek, and the Canadian Pacific Railway picked the spot as its western terminus. Young men flooded in for the Fraser Canyon rush, heeding the famous motto "Get in, get yer gold, and get out." Several local penny exchanges provided an easy means of raising capital for these schemes, and in 1907, the omnibus Vancouver Stock Exchange was chartered.

The frauds began almost immediately. Promoters hawked shares in whaling ships, timber stands, and bird guano fertilizer companies. Minerals were always the star attraction, though—the notion of buried treasure is a durable opiate. Otherwise prudent men could be charmed into thinking that a faraway patch of ground *might* be the one concealing a vein of gold or silver. It was more mysterious than textile factories or hotel chains—more subject to huge payoffs and crushing disappointments, and therefore more attractive to a certain risk-drunk personality. Mark Twain's definition of a mine as "a hole in the ground with a liar at the top" was never truer than in the early days of the Vancouver exchange. And the rich wilderness of British Columbia had made it possible.

"The Canadian economy was started on ventures—timber, trapping, and minerals—and Canadian exchanges were built around these industries," said Paul MacKenzie, the president of Red Hill Energy. "We've always had risk money going into minerals. Canadian investors still have a thirst for this."

Enough of the offerings hit big to justify thousands of lemons. In 1964, a company with the unfortunate name of Pyramid was selling at 35 cents a share, advertising some lead and zinc claims in the Northwest Territories in a joint venture with the mining giant Teck Cominco. The crew had a drill rig out at a frozen spot called Pine Point, and the foreman was supposed to relay encouraging news over the radio with the code words 3 *a.m.* in hopes of confusing potential eavesdroppers. He got on the radio with a cheerful tone of voice one day, obviously drunk, and his supervisors kept interrupting him to ask him what time it was. "I don't give a fuck what time it is," he slurred into the microphone, "we're all fucking millionaires!"

Pyramid's drill had intersected a convincing amount of zinc and lead, and the stock went up to $14 by the end of the next day. This success only fueled a host of imitators, who rushed to stake everything surrounding Pine Point and sell it as the next big thing. This is known in local parlance as an area play.

Things had gotten so bad by 1989 that *Forbes* magazine was moved to call Vancouver the "Scam Capital of the World" and to note that all the unregulated capital flowing through its exchange had made it a laundry for mobsters. "Each year it sucks billions of dollars out of legitimate markets by inducing dupes in North America and Europe to invest in mysterious outfits making hydrodouches, computerized golf courses, and airborne farm equipment," said the magazine, adding, "Nobody blinks or even chuckles."

Reforms were promised, but delivered only after a disgraceful episode on the more respected Toronto Stock Exchange. In 1997, a company called Bre-X started spreading rumors of a huge gold discovery on the island of Borneo. There was said to be a confirmed deposit of seventy million ounces in the jungle, with core samples to prove it. Investors rushed to buy shares in Bre-X, which quickly acquired a market capitalization of $4.4 billion and a partnership with two majors, as well as an alliance with a company controlled by a son of the corrupt Indonesian dictator Suharto. Canada's brokerage houses responded with glee, and the stock price climbed even higher. "They seemed to have teamed up with a partner with high standing," a mining analyst with the firm Lévesque Beaubien Geoffrion told a reporter.

The truth started to come out after the core samples were taken to an independent lab. They had been sprinkled with gold shavings, a practice known as salting and one of the basic scams of the mining trade. The chief geologist for Bre-X began to say he was haunted by "evil spirits," and then he tried to kill himself by drinking a bottle of cough syrup and lying facedown in the bathtub. A few days later, he fell eight hundred feet from a helicopter into the Indonesian jungle. His body was found after four days, decayed and partially eaten by wild pigs. No autopsy could be performed, and rumors circulated that he had not committed suicide, as the official verdict said, but had been pushed. The founder of Bre-X died of a stroke in the Bahamas the next year. In what many consider an embarrassment for the Royal Canadian Mounted Police, none of the surviving officials was convicted of a crime.

The atmosphere in Vancouver is somewhat less sleazy today, though the regulation is still light and the odds of any one claim yielding hard revenue are just as astronomical. Canadian mining companies are now required to file a National Instrument 43-101, a certificate signed by a licensed geologist, after they make any public statements about the potential of one of their claims. This information must also be in the press releases sent out to bump a stock; however, the data can be impenetrable to nonexperts, full of statistics and code names and laden with postdoctoral language. Punishment is light for a company that tells outright lies; the worst thing that usually happens is a temporary halt in trading until corrections can be issued.

The most typical scam perpetuated in Vancouver today is no scam at all by strictest legal terms, though it is no less dishonest, and no less dependent on the naïveté or greed of the investors. A junior company of two or three friends simply picks a region somewhere on the globe where a small amount of uranium has been detected (not hard to do: The stuff is a common element in the earth's crust) and legally stakes a few hectares. Then a minimal amount of geophysical work is done, a technically truthful press release issued, and a Web site put up. The stock gets listed, and the project is hyped to brokers and the newsletter writers who cover both the penny stocks on the Toronto Venture Exchange and the even less well-regulated Over-the-Counter and Pink Sheet exchanges in the United States. The property is never drilled for core samples. That's too expensive. And it will likely provide hard evidence that the site is a dog—hence the old saying "If you drill it, you kill it." The company makes scrupulously honest statements about absolutely everything except what they know to be true in their hearts: That not a single ounce of uranium will ever appear.

Given the dismal climate in the Vancouver markets, what kind of person would put his or her income into such a known meat grinder?

The average investor is a middle-aged male with a few thousand dollars who feels that he hasn't yet achieved the financial success he deserves and that uranium is a way to catch up quickly. Few know how to wade through the numbing technical language of a 43-101, but there are newsletters and brokers to guide those decisions and, in any case, *the story* has always been more important than the science. These stocks cost pennies per share, they rise fast, crash overnight, and provide entertainment far beyond their face worth. One junior president compared it to buying a raffle ticket. "Losses are limited," he said. "And thrills are high."

These snowball-chance plays offer the investor something even more important: the intangible value of respect; of being someone who matters. "A little guy can put down five to ten grand and that's a significant position," one investor relations man told me. "He's a player. They will take his phone calls." Such a thing would not be possible at Coca-Cola or Google. Uranium thus offers the average lowballer a chance to feel like an earl of finance, the same way that any man who walks into a strip club is immediately lavished with attention and conversation.

Talk of the nuclear renaissance made uranium speculation the hot thing in Vancouver in the fall of 2007, with a market capitalization estimated at $250 million for the juniors alone. The number of exploratory projects was said to be ten times those active during the dog days of the eighties.

Not all of these companies are disingenuous. Quite a few are willing to do the drilling—to use what Jim Cambon calls the rotary truth machine—and operate in the good-faith hope that a multinational will consider their findings worthy of a joint venture or an acquisition. The only way to do that is to walk the path of all the fakers and make the dirt look as good as possible. Juniors previously occupied with diamonds or copper or molybdenum started adding uranium to their suite of minerals just to be safe. A number of them changed their company names to incorporate the word *uranium*.

"In a few short months," exulted an investment guide, "uranium has become one of the most sought-after commodities since the Romans minted gold coins." This reborn market in radioactivity floats on a raft of hope.

That hope has now extended even into Yemen, home of the would-be nuclear reactor on the shores of the Red Sea. A small Canadian company announced the staking of a possible deposit some miles to the southwest of Sana'a. The evidence was visible on the old maps of an airborne radiometric survey done in 1992 by a British company. Gamma rays were detected. It might have been a surface layer of potassium or a similar metal that throws off a radioactive signature, but it could well have been uranium.

I called on the company's office in Yemen, a gray gingerbread-style house behind a high gate of sheet metal. The geologist was an older man with blue ink stains on his shirt. He invited me in for tea and showed me the place on the map where, he assured me, the exploration would be continuing.

"We would like to see if there's something in the ground," he told me. "*Inshallah*, we will begin work in the next two or three weeks. We are just waiting on some instruments."

There was a hole in the ground, a perfect circle, a dark eye bleeding small pools of green liquid from the edges. A drill had just been pulled from it.

It was sunset on the edge of the Gobi Desert in southern Mongolia, and I stood near the hole with D. Enkhbayasgalan, a twenty-five-year-old geologist (almost all Mongolians have only a single name but use the initial of their father's name on formal occasions). He pointed east across the gravel hardpan toward a straight line of similar drill holes marching off into the distance, each about the length of a dollar bill in diameter and each one intersecting uranium that began about forty yards down. The past week's work had revealed that the ore body was about two miles long and was shaped like a salamander.

"If we get lucky," said Enkhbayasgalan, "we'll even be able to see the crystals under a microscope." He wore a baby blue sweatshirt with the words COLORADO USA.

The drill hole was part of a series—a "profile"—called East Haraat, which referred to a distinctive rock outcropping that had functioned as a lookout for Genghis Khan's horsemen in the thirteenth century. From the top of it, there was a commanding view of the dry plains to the south, the grass and sky two halves of a sphere that seemed to encompass all. There was now a uranium camp at the base of the hill, run by a company from Canada named Denison Mines, with a water tower, a generator, three wooden cabins in the dacha style, a metal-sided horse corral, and a line of eerie futurist streetlights that glowed chlorine green at night. The camp had been built in the midst of this spectacular isolation by construction crews from the Soviet Union in the 1970s, when Mongolia was a loyal client state and a reliable source of uranium. They had not managed to develop a mine in this part of the Gobi, but they drilled thousands of holes, and their old radiometric logs inscribed with Cyrillic lettering became valuable after the price of uranium started to climb and Western companies looked to Mongolia as the next hot frontier, with a pliant government and the peal of wealth lying under the grass.

Among the usual legalized frauds from Canada were several well-capitalized outfits that had every intention of resurrecting the Russian uranium archipelago and railing the ore to China. But the government became hesitant. In 2006, the Mongolian parliament rescinded a generous royalty schedule that had been passed in the days after the collapse of Communism, when hard currency was scarcer than it is today. Mongolia is still trying to decide just how much of its uranium it wants to give away and at what cost. The lessons of the "resource curse" are remembered here, and the government does not want to turn itself into another Niger—a slave to its own geology.

Mongolia has a heritage of plunder and abuse, dealt from its own hands at the height of its thirteenth-century military glory but received from its neighbors ever since. The nation is a landlocked plateau wedged between two great powers, Russia and China, which have treated it as a stepchild and a cash cow ever since the collapse of the Mongol empire. China finally left in 1911, when the Qing dynasty fell apart, and the Soviet Union moved in ten years later, setting up a puppet government and launching violent purges of the Buddhist monasteries, which they viewed as rivals for power. They built instant towns with big drafty buildings and ghastly apartment rectangles at calculated spots in the grasslands and forced a nomadic people to take up a more European-style life of wages and timetables. The capital city was located at a lonesome river crossing called Ulaanbaatar—the name means "Red Hero," for a local party hack—and decked with town houses, wide triumphant avenues, vodka bars, a ceremonial Parliament, a tourist hotel with bugged rooms. Ulaanbaatar was known as the world's coldest capital, a hardship posting for diplomats, where winter winds howled all night long through barren concrete plazas. On the city fringes were haphazard arrangements of gers—the traditional circular tents of birch poles and thick wool that can be erected and reerected in a matter of hours and that have been a housing staple here for thousands of years.

The Soviets could not eliminate the pastoral life, and its economy of cashmere and sheep's milk, but they could wipe away traces of regional pride as embodied by the figure of Genghis Khan, who was never mentioned in schoolbooks or honored with a statue, though he had represented the pinnacle of Mongol glory, organizing rival tribes into deadly phalanxes that marched out of the grasslands in 1211 and started taking Chinese cities. With fast horses and a deft series of alliances, they moved into the breadbaskets of Persia and extended their reach to the Caspian Sea, becoming a world military superpower. Some of the empire's people are believed to have walked much earlier across the ice-choked Bering Strait to live in Alaska, where yet another offshoot* settled the gorgeous (and uranium-rich) lands of the American Southwest and became known as the Navajo.

Khan died in 1227 and was buried in an unmarked grave in the hills northeast of the present-day site of Ulaanbaatar; his pallbearers were all slaughtered so they could never reveal the spot. Khan's sons and grandsons extended their patriarch's domination, reaching the foothills outside Vienna and sacking the scholarly center of Baghdad, spelling the beginning of the end of Islamic cultural dominance. Hundreds of thousands of people were efficiently put to the sword. The irreplaceable contents of the caliph's libraries were tossed in the Tigris River, said by chroniclers—with a touch more color than accuracy—to have run black with ink after it stopped running red with blood. Hearing of these disasters, the monk Matthew Paris was moved to call the Mongols a "detestable nation of Satan" and harbingers of the coming apocalypse. But they could not maintain their rule by force. Infighting and poor administration caused the empire to fall apart at the beginning of the fifteenth century. At its apex, it had covered more than four times the lands conquered by Alexander the Great.

A well-worn joke today is to refer to a strongly conservative person as being "to the right of Genghis Khan," but the joke gets it wrong: Khan was, if anything, a liberal by modern definitions; he instituted a government system of record keeping, put the Mongolian language into writing, outlawed the kidnapping of women, assured diplomatic immunity to his neighbors, established an independent judiciary, and levied stiff taxes on the lands he conquered. As the historian Timothy May has noted,

---

* An unproven and somewhat controversial theory: Some words, customs, and beliefs are shared by both cultures. The Navajo hogan is said to be a New World version of the *ger*.

he also instituted a welfare system for widows. Khan was also known for his sexual appetite: DNA tests reveal that about one-sixteenth of the population of eastern Asia is genetically descended from a single person, believed to be him.

The people of Mongolia held on to their love of open vistas and rambling even while under the Soviet occupation. About 40 percent of the population lives untethered to any city, raising horses, camels, sheep, and dogs in a cycle of grazing encampments that revolve with the seasons. When approaching a *ger*, a visitor is supposed to yell *"Noho hori,"* which means "Hold the dog!" This is a version of hello. It is not uncommon to see motorcycles parked outside a *ger* in the middle of nowhere, though a visitor is always welcomed with tea and curdled sheep's cheese. Hospitality is a social necessity in a country where the temperatures routinely fall below minus 20 degrees Fahrenheit in the winters.

The big rolling prairies are reminiscent of eastern Wyoming in their scope and blankness and annihilating reach. This is a land that does not easily show its scars, and the few cities the Russians managed to leave have a tumbledown aspect. Scavengers have been at work on the ruins of the old military installations, and even the concrete walls are coming down, fading into the grass like the remains of a Roman garrison. Nothing seems permanent.

I went out again to the edge of the Gobi with a lanky operations manager who peppered his speech with *bloody* and *fuck* and rattled his Land Cruiser at high speed down a washboarded path that paralleled the Trans-Mongolian Railway. The asphalt had given way to dirt as the grasses had thinned away, and the horizon flattened into a long dun-colored plain the deeper we went into the desert. We pulled over for a lunch of sandwiches and coffee, shielded from the cold breeze by a mound of weedy earth, and watched a man clad in a blue robe walk past us silently and toward the next rise, some five miles off. We were still watching him half an hour later, a tiny blue dot bobbing on the far horizon, next to the bumpy dirt track.

"This is the main road to Sainshand?" I asked, naming a town I had seen on the map.

"Hell," he told me. "This is the main road to *China*."

The primitive condition of this highway to Beijing had suited the

Russians, who were wary of creating an easy invasion corridor for their Chinese rivals to the south. The area closest to the border had been seeded with land mines, which the camels kept setting off. The railroad tracks in China are still different widths from those on the Trans-Mongolian, and freight cars must be lifted onto dollies for transfer at the border. This is the awkward point of juncture for two civilizations that have been in contact for more than three thousand years.

Another legacy of the Russian occupation can be found a few hundred miles northeast from the presumed spot of Genghis Khan's unmarked grave, in a series of low waves of land called the Saddle Hills. There was once a secret uranium city here, a place named Mardai that appeared on no maps but was home to thirteen thousand workers and a branch of the GUM state department store. The local herdsmen were warned that the region was cursed, full of poisoned lakes and sheep with only three legs.

Almost nothing but concrete slabs and broken glass is left today. The city has been vandalized to the point of nonexistence, the shell of its downtown sledgehammered to chips for the sake of the rebar at the core. The scrap metal is valuable, and deconstruction can be as profitable as construction. But in its heyday in the 1980s, Mardai was the pride of the Soviet military, classified even within parts of the Kremlin and exempt from the usual idiocies of five-year plans. Its managers were told to drill as deep into the Paleozoic as they needed. Their only mandate was to develop a strategic reserve of uranium for national emergencies. After President Ronald Reagan started making bellicose speeches in 1982, the mines went underground. The Soviet engineers sank a total of eight shafts, intersecting dark veins of pitchblende every place they drilled.

"Thank God they threw that extra money at it!" said Gerald Harper, the vice president of exploration at Vancouver-based Western Prospector Group. "There are sixteen kilometers of ventilation pipes down there that still work. It was well-done work with high-quality materials."

Western Prospector now owns a good portion of the once-secret city and plans to start selling off the Red Army's old uranium for reactor fuel. The company has built a mining camp big enough for two hundred workers and hopes to be busy exploiting the Russian-built shafts by 2010. The camp will be dismantled when the uranium is gone, as efficiently as Mardai disappeared into the grass. "We do not want the hassle of running a town site, no schools or hospitals," Harper told me. "The last thing we want is to be (a) a social service agency and (b) to have that liability when the mine closes down." The workers will likely be bussed in, he said.

Western Prospector had acquired its part of the secret city from an old friend of Bob Adams's, a bald-headed veteran of the Texas uranium fields named Wallace Mays, who had spotted a gleaming opportunity in Mongolia shortly after the Russians departed. He made a deal to acquire a one-third interest in the site, and went on to claim more than half, though he would eventually lose most of it, reportedly due to financial troubles. During his many trips to Ulaanbaatar, Mays—in his seventies—met and married a local woman named Hulan nearly half a century younger than he. Mays told me over drinks at a Toronto hotel that the culture of horsemanship and the free range shared by both Texas and Mongolia were critical to their bond. He is known in the uranium business as one of its old-time cowboys—brash and confident. Mays has often claimed to have invented the best modern system for in situ mining and got rich by opening several mines in Kazakhstan. "I didn't develop them for the money," he told me later. "I developed them for recognition."

His plan to exploit Mardai was thrown into doubt in August 2007 after the government suddenly revoked the exploration licenses. The decision was reversed after a flurry of protest from Vancouver, but the flap exposed one of the strong-arm tactics that the government might reserve for itself if it feels the uranium frenzy becomes too destructive. The Mongolian Parliament granted itself the right to acquire up to 50 percent of a mine deemed to be of "strategic" value. Its logic was that local shell companies had done the hard work of exploration during Soviet times, and the nation was thereby due a fair share. But the problem is that nobody in Ulaanbaatar has yet formulated a definition of "strategic." Some of the more hard-core Socialist elements of the government have called for nationalizing the mines.

For better or worse, the initial tone in Mongolia has been set by Robert Friedland, a man described with awe in Vancouver as "the greatest stock promoter of all time." He is also known by environmentalists, less respectfully, as "Toxic Bob," because of an incident at Summitville, Colorado, in which the state and federal governments were stuck with a $200 million cleanup caused partly by a cyanide leak from the premises of a gold company of which he was CEO (he has denied the allegation). Friedland is a long-faced graduate of Oregon's Reed College, a onetime student activist who studied Buddhism and eventually found a career in the Vancouver financial markets. He is now the executive chairman of Ivanhoe Mines, a company that claimed an area the size of Connecticut in the Gobi Desert, encompassing what it says is the largest copper and gold deposit in the world, at a place called Oyu Tolgoi, close to the Chinese border. It is expected to reap $2 billion a year, more than twice the current gross national product of Mongolia.

Friedland has promised to employ thousands of locals and contribute liberally to social causes within Mongolia, but protesters have burned him in effigy outside Parliament, and he remains a controversial figure. Friedland only fueled the controversy with a 2005 speech to a group of investors at the Royal West Hotel in Tampa, Florida (a performance now known as the T-Shirt Speech), in which he called Mongolia "the hottest exploration venue on planet Earth," which he hoped to build into a "mining country like Chile." There was plenty of land around for waste dumps, and Mongolia was close enough to the Chinese border to feed copper to the Chinese. The most notorious part of the speech, however, came as he was describing the block-caving method by which the minerals would be cheaply liberated from the host rock. "You're in the T-shirt business," he said, "you're making T-shirts for five bucks and selling them for one hundred dollars. That is a robust margin." The line was heard as an allusion to sweatshops, and it did not go over well in Ulaanbaatar, where politicians were quickly furnished with copies of his words. But Mongolia's weak economy subsists on mining—minerals of various sorts represent 70 percent of all exports, ahead of cashmere wool—and the government was looking for ways to appease populist sentiments while preserving a reputation as an investment-friendly place.

I went to see D. Javkhlanbold, the head of the geological department at the agency that had pulled the licenses at Mardai. He was a man in his twenties with a hard handshake and a colorful necktie with a map of the world as its pattern. I asked him about the current discussion over which uranium field would eventually be considered "strategic," and therefore open to heavy state participation. He told me that much of the original exploration had been done by Mongolian companies acting under Soviet duress. The minerals therefore belonged to them, too. Those fields judged the richest would be most likely to be partially subsumed, he said.

There was a hope that Mongolia could one day build its own atomic power plants, he said. They would be fueled by local uranium. The nuclear renaissance should also benefit those places that provide its seed material, much as the Belgian Congo had been rewarded with a nuclear reactor for its role in helping to exploit the uranium treasure at Shinkolobwe. But this hope, admitted Javkhlanbold, was far off. There were still a lot of questions that didn't have any answers.

"Right now," he told me, "there's not a lot of understanding of uranium."

9

LEGACY

In the rush to put old fields back into production, one place was not forgotten.

A company called Brinkley Mining signed an agreement in 2007 with the Democratic Republic of Congo for the exploitation of five known deposits of uranium. A priority target was the old pit at Shinkolobwe, home of the homicidal ghost Madame Kipese and the birthplace of Fat Man, the seam in the earth that refused to stay closed, where the supply of uranium seemed to be without end.

As a bonus to the Congo—and perhaps a sign of eagerness—the London-based Brinkley also promised to help fix the broken nuclear plant in Kinshasa and install radiation detectors at the borders to keep Shinkolobwe's stolen uranium from being smuggled out of the country.

But the deal was quashed, as so many things are in the Congo, by corruption. Police arrested two of Kinshasa's top nuclear officials and accused them of conspiring in a plot to export uranium without a permit. A deputy minister canceled Brinkley's leases. "Uranium is a reserved mineral," he told a reporter. "We want to leave it for future generations." The Congolese government then turned around and sold the prospecting rights to the French nuclear giant Areva.

The mystery of the illicit buying at Shinkolobwe, meanwhile, was never solved. There was no accounting of how many bags of uranium might have left the Congo under a truck tarp and smuggled to places unknown. Western diplomats suggested that purchasing agents for Iran, or perhaps even North Korea, may have been the ultimate buyers. It could also have been elements of the A. Q. Khan sales network, or terrorists looking for shrapnel in a dirty bomb.

A more banal possibility, and one more likely than any of these, is that it was simply hoarded up by a speculator waiting for a buyer to come forward, much as Edgar Sengier had kept his own barrels from Shinkolobwe in a vegetable oil plant on Staten Island in 1940, patiently waiting for a visit from the U.S. Army.

The uranium could not be fashioned into a weapon by itself, but it might be useful to a state with nuclear ambitions. The purloined ore could be fed into a graphite-moderated heavy-water reactor, such as those now located at Khushab in Pakistan or Arak in Iran or even Cirrus in India, which can run on natural uranium. Though this would be a cheaper path to a bomb than conventional enrichment, such a scheme would require a campus that would be hard to conceal from spy satellites. At a minimum, there would have to be a yellowcake mill, a fuel fabrication plant, a reactor, and a sophisticated reprocessing shop with glove boxes, precision gear, and tubs of nitric acid. The Israelis built their bomb in the 1960s using just this method, but they had to hire a French contractor and bury their equipment six stories deep. The supply of raw uranium ore would have to exceed six tons at a minimum.

A nation which attempted such a cut-rate Manhattan Project would face formidable barriers. It would need millions of dollars and the unpredictable factor of luck. But seventy years of history has shown that nations are willing to make extraordinary sacrifices and challenge long odds to enter the club of the privileged. Nine nations have made the journey thus far. Pakistan's Z. A. Bhutto had once promised his countrymen would "eat grass" for the sake of atomic potency, and the furtive ingenuity of A. Q. Khan matched the sense of that determination, if not the colorful actuality. The Soviet Union turned entire mountain ranges into gulags for the sake of pulling even with America in the arms race; America itself had made a massive wager on an uncertain hypothesis in a time of war, and had later sent at least six hundred—probably thousands—to cancerous death for the sake of a commanding nuclear edge. Israel, South Africa, India, and North Korea have all followed in their own pursuit of the uranium golem, spending huge amounts of capital and risking war.

Iran may be next. The first century of our experience is not yet over. History is long.

If a speculator did take a flier in black-market uranium from Shinkolobwe, it would not be difficult to hide it from prying eyes. It could be barreled and stacked in an obscure corner of an industrial yard. It could be stored in a row of tin sheds in a forest. For that matter, it could simply be piled out in the open air, with the perfect disguise as a gravel heap. Rain or snow will not harm it. And its fissile potency will not substantially diminish until approximately seven hundred million years have passed.

The uranium from Shinkolobwe is, in fact, so freakishly powerful, that even its waste has attained the status of legend among scientists trained to deal with such matters.

The waste from America's first atomic bomb still has its own special code name—"K-65"—and it continues to pose a unique environmental menace without any clear solutions in sight. How deadly is this material? Remnants from typical uranium from the southwestern United States give a radioactive signature of about forty picocuries per gram, about ten times the amount of picocuries per liter of air that is considered safe for humans to breathe. The Shinkolobwe remnants, by contrast, emit a stunning 520,000 picocuries per gram. The effect across the register is like comparing a housefly to a moon rocket. But this awesomely powerful waste had been treated haphazardly, even casually, by the private corporations who had played a crucial role in the building of America's first atomic bomb.

The ring-and-plug heart of the Hiroshima bomb had been fashioned at the Linde Air Products and Electro Metallurgical plant in the smokestack flatlands just east of Niagara Falls, New York. This was a natural place for a heavy industrial project such as the bomb. The relentless tide of water draining from the Great Lakes at Niagara Falls was already one of the country's great chemical manufacturing hives, a source of cheap electricity and labor—and the eventual home to one of the more notorious environmental scandals of the twentieth century with a direct link to uranium.

It had begun when a promoter with the soft name of William T. Love started dreaming of a multilevel canal that would link the Niagara River above and below the falls, essentially creating a second river. The descending canal water would generate electricity and help pay for itself, Love promised his investors. He also touted a nearby settlement of homes and factories called Model City. Local burghers worried the diverted flow would shrink their falls—a kitschy honeymoon destination and, as the old joke went, "the second-biggest disappointment of a bride's wedding night"—but were spared from taking action against Love when the Panic of 1893 wiped out his joint-stock company. He only managed to dig out a single mile of his canal, which became a swimming hole for local children.

By the 1940s, Love Canal had become a trash dump and was eventually used as a disposal site by the nearby Hooker Electrochemical Company, whose most important customer was the Manhattan Project. The products they provided were lubricants and fluorine, crucial to the formation of the uranium hexafluoride gas that was run through the pipe mazes at Oak Ridge. Hooker also was host to a secret manufacturing facility for ceramic parts for the atomic bomb. The code name for this place was "P-45," and it was considered so crucial that Niagara Falls was one of only six locations in the country to host a field office for the Manhattan Project.

Working together, Hooker and the U.S. Army dumped thousands of barrels of waste in the Love Canal, often in secret and at night. Local observers later said it was commonplace to see green army trucks with white stars on their doors driving up to the canal at midnight and crews of armed soldiers unloading barrels full of chemicals. This went on until 1952, when the trench was covered with a thin layer of topsoil and sold—albeit reluctantly—to the Niagara Falls School Board for $1 the next year as a site for a new elementary school to handle the crush of young families seeking a low-cost tract house in suburbia. A working-class neighborhood was put up around the 99th Street School and it wasn't long before rainwater eroded the topsoil from the top of the old canal, exposing the rotting tops of steel barrels and creating puddles for the children to splash in. A persistent, eerie smell of acid began to pervade the new neighborhood. Residents began to see oozing sludge creeping in between the cracks in their basement walls. Sections of the old landfill intermittently caught fire. The plastic in aboveground swimming pools dissolved. Children always seemed to be sick. Those who weren't came home from school with burn marks on their faces and arms from playing in the grass.

Then the birth defects began to appear. One infant's teeth started to appear in double rows. Another came out of the womb with three ears on its skull. Others were miscarried or born retarded. In 1978, the residents finally got the attention of President Jimmy Carter, who helped evacuate the area and effectively declared it a disaster area. Love Canal had become a marker of seventies environmental dread, even though its disease could be directly traced to the dawn of the atomic age.

Twenty miles away from the canal, near the site of Love's failed Model City, another historic malignancy was festering inside a basement at the site of the old Lake Ontario Ordnance Works. This had been a famously productive manufacturing plant for explosives, active between the years of 1942–43, when thousands of tons of its TNT was packed into artillery shells and used to kill Germans and Japanese on the battlefields. To confuse possible saboteurs and Axis spies, many of the buildings were disguised to look like livestock barns and the guards patrolling the edges wore farmer's overalls. Near the end of the war, when nearby Linde Air needed a place to get rid of its cuttings from the African uranium, the 191-acre explosives plant offered a seemingly ideal spot. It was already a condemned federal dominion for one thing, tucked off in the deciduous forest out of the sight lines of nearby farms and houses. And there were plenty of abandoned industrial structures in which to lock away angry trash. The atomic waste—more than two hundred dump trucks of it— was sealed into drums and stacked inside a concrete silo that had once held water. It was never designed to hold anything radioactive.

The government knew this was a bad idea, but did not say anything to the public. According to a June 2, 1956, memo from S. R. Sapirie of the Atomic Energy Commission, the "choice of site hinged more on availability rather than any unique features making it suitable for such storage." The uranium barrels were also "extremely contaminated" and the leakage "has apparently resulted in the spread of some measureable activity [*sic*] to some distance outside the reservation."

What this meant was that the groundwater near Lake Ontario had become radioactive, but nothing was done about it for more than twenty years.

After the Love Canal fiasco broke wide in the early 1980s, and under pressure from New York State of Assembly, the U.S. Army Corps of Engineers decided their uranium waste had to be dealt with in some fashion. And so the Shinkolobwe ore was scooped out of its steel barrels (themselves so radioactive they had to be shipped to Painesville, Ohio, for special disposal) and a piping system made out of firehoses scrounged from local fire departments was built to suck the waste in a darkwater slurry over to the basement of a nearby dynamite warehouse once called Building 411. There the waste was molded into a neat rectangle by a bulldozer. This machine finished the job so hot that federal officials simply buried it alongside the uranium. The firehoses, too, were thrown in. Also interred there were several dozen dead beagles, which had been used in radiological experiments at the University of Rochester. This "Interim Waste Containment Structure," as the army called it, was covered with thirty feet of soil and a blanket of bluegrass was planted on top. This was supposed to be a temporary solution until a better idea came along. But nothing has been done since.

"They had to put it *somewhere* and I'm here to make sure everything is maintained and stabilized," said Dennis Rimer, the site superintendent. "It's prettier than my yard will ever be. It's like a park to me."

I met with him on a June morning inside a rehabbed office at the old dynamite factory, at what is now called the Niagara Falls Storage Site of the U.S. Army Corps of Engineers. Its primary mission, with an annual budget of a half million dollars, is to watch the grave of the Shinkolobwe leftovers—a geometrically precise mound that is one-fifth of a mile long and covers ten acres. Dennis wore a red polo shirt with an embroidered lizard and the words "St. Maartens" on the chest, a memento of a cruise ship vacation the year before. Clipped to the open neck near the buttons was a slim gamma ray detector which had never once registered, he said, adding, "There's lower background radiation here than in some people's yards."

Yet there are limits to its stability. The basement of the warehouse was never intended to be a final solution and even the army admits the waste will start leaking into the groundwater at some point, sending fingers of radium and uranium toward the Great Lakes. They peg that date more than a century in the future.

That estimate does not satisfy Timothy Henderson, who is the president of a group called Residents Organized for the Lewiston-Porter Environment. He works the graveyard shift at a local water-treatment plant and is worried that the bomb wastes have been contaminating the aquifer in the new site all along.

There is a precedent for this, as well as the official half-truths and corporate disinformation that have characterized the legacy of atomic bomb residues in the region. The Linde company, for example, found it easiest to dump their old uranium sludge directly into underground wells where it could not be blamed on them. "Radioactive isotopes eventually found their way into local creeks," wrote the cultural historian Ginger Strand, "and ultimately the nearby Niagara River, which means that uranium from building the Bomb went over Niagara Falls." State data shows that breast cancer in the immediate vicinity of the Shinkolobwe waste dump happens 49 percent more than the average and prostate cancer at an even higher rate. Tying such health patterns directly to the Manhattan Project fragments isn't yet possible from an epidemiologic perspective (the whole area is awash in contaminants), though Henderson finds the nuclear evidence compelling.

"It's stuff that never should have been brought here in the first place that's destroyed this whole area," he said. "They knew the water table was not appropriate for burying it here. It'd be a lot cheaper to move it. We don't want to be the ultimate victims of history's ultimate weapon," he said.

The army's project engineer, Michelle Rhodes, drove me around the site in a motorized greenskeeper's buggy. We went past a third-growth forest of ash, maple, and elm along crumbling roads that had linked the vanished dynamite buildings to one another. A set of concrete cradles that had once suspended vats of nitric acid above railroad cars was overgrown with moss and tree branches, receding into the forest. It looked like a postcard view of a crumbling Roman aqueduct. To the east was an active land-fill run by a private concern called the Modern Corporation—it housed a good percentage of the current garbage outflow for the suburbs of Buffalo.

As we bounced along, Rhodes told me she had held a job designing microscopes before coming to work for the nuclear dump site. A few years ago, she was out taking gamma ray samples on top of the mound when she came across a piece of obsidian rock jutting out of the topsoil, looking like nothing so much as a chip of furnace coal. Its radiation readings were spectacular: here was a chipping from the Shinkolobwe pit that had somehow migrated upward through this pile of New York soil.

"I remember holding it for a while," said Rhodes, "and thinking to myself, 'Wow. So this is what it looks like. This is a piece of what was in the silo.'"

She let us through the cyclone fence that surrounds the Interim Waste Control Structure and drove us up on top of a grassy tumulus. Dennis Rimer had been right: on the surface, it looked as benign and manicured as a golf course fairway. One could imagine a decent game of ultimate Frisbee or a picnic up here on top of this radioactive ground shipped halfway around the world. The actual ore from the Congo lay thirty feet below us, capped with a layer of clay. Rhodes told me the mound occasionally reminded her of a burial mound near the shores of Lake Ontario from the Hopewell Indians who had lived here in the first century. Archaeologists know that it was for the interment of corpses but were unable to discern its spiritual importance to the vanished culture.

"This was never really designed to be permanent," she said of the uranium dump. "The problem with K-65 is that there's no landfill that wants to take it. It is different than anything we have ever dealt with."

The only real solution to the Shinkolobwe problem in America, concluded an official from the New York Department of Environmental Conservation, was "assurance of over ten thousand years of isolation from mankind."

The chain reaction of uranium—that trick of physics uncloaked in the 1930s—had first been detected because of its waste products, and it is the waste that remains long after the electromagnetic bonds have been severed. Shards of cesium, strontium, plutonium, angry and newborn, fallen away from the original nucleus like slices of an apple. Enrico Fermi saw this dangerous mess in a Roman laboratory in 1934 and it signified the first understanding of a hard truth. Uranium cannot fission without creating waste, just as there can be no digestion of food without feces. And still, the problem of nuclear waste has yet to be solved by any nation.

There is a person on the payroll of the American government whose job it is to think about ways to warn future civilizations (this includes possible extraterrestrial visitors) about the presence of buried nuclear waste. The U.S. Department of Energy feels an obligation to extend a warning ten thousand years into the future that "Here lies poison and death." The survival of the English language cannot be assured, just as nobody today can speak Sumerian. A stone marker that reads "Warning," or shows a graphic of skull and crossbones—for who knows if the intruders will even possess skulls as we know them?—will not be sufficient. This type of etching must transcend English or any signified language. It must be a kind of architecture or hieroglyph that has no language.

The man paid to contemplate "deep time" and its relation to nuclear waste is an avuncular lapsed Mormon named Abraham Van Luik and he occupies a windowless room inside the suburban Las Vegas office of the Yucca Mountain Project, the plan approved by Congress to hollow out a mountain in the Nevada desert and fill it up with spent fuel from the nation's 103 nuclear plants. Van Luik shows me a rendering of the monument system that he likes best—a set of three warning monuments shaped like fan blades that would look like the classic three-winged nuclear symbol when seen from the air. Made from yellow granite or basalt, there would be four of these erected at the perimeter and house documents inside that explain the exact contents of the waste dump below.

"You do not have to make those markers ugly and repulsive," Van Luik told me. "Ours is going to be a little more descriptive—it's hoping to get an intellectual message across. Each one of these will be a little temple you can walk into."

These would be temples not of the religious sort, he assures me, but of the scientific. The carvings on the outside would be in seven languages and include signs and symbols intending to convey danger in a way that transcends language. Notes the U.S. Department of Energy: "They would be designed to be unnatural looking so they would draw attention, but not be misconstrued as memorials of honor." Materials must be of excellent quality, because these monuments would have to approach the durability of the plutonium inside, which has—relative to the other waste—a relatively brief half-life of twenty-four thousand years, about ten times as long as the Parthenon has stood in Athens.

Yucca Mountain is a ridge of volcanic rock known as "tuff," zigzagging north-south about ninety-five miles to the northwest of Las Vegas. Barely seven inches of rain falls here per year. There are small concave openings resembling caves near the crown of rocky cliffs at its top that were used by hunter-gatherer peoples who roamed the Great Basin and neighboring Death Valley twelve thousand years ago. They left a constellation of stone tool points to mark their presence. Everything else: baskets, wickiup, clothing, garbage, has long since disintegrated in the hot desert air. The slopes are lightly covered with obsidian rock and bur sage. Anglo settlers at the turn of the twentieth century judged it close to worthless; it held no gold or silver and the soil in the basin was far too arid to grow anything. Small pockets of clay and mercury were mined out and then deserted. The nearest city of note, Rhyolite, rose from nothing and fell within the space of fifteen years, becoming a ghost town by 1917. It left behind a few limestone shells on its main street, an abandoned row of prostitute cribs to the side, and a defunct stagecoach road across the southern flank of Yucca Mountain.

In 2002, after twenty years of study and $4 billion (double the price tag of the entire Manhattan Project) spent on tests and research, the U.S. Congress identified this place as the chosen eternal vault for 154 million pounds of old American uranium, most of it still inside assembly rods that look like beds of long, slim nails. Competing sites in Texas and Washington state were dismissed, and boring crews began digging a three mile tunnel into the mountain to further examine the way that water seeped through the volcanic fissures and serve as an eventual subterranean railway for the shelving of waste canisters inside different alcoves.

The selection of Yucca Mountain had as much to do with psychology of landscape as it did science. The Nevada Test Site, the arid reserve where so many weapons had been detonated in the 1950s, was nearby and there was a general perception that the region was already nuclearized to a hopeless degree and that more hazardous castoffs couldn't hurt such a blasted plutonic moonscape in any case. It embodied the attitude once characterized by author David Darlington of "the desert-as-dump"—that is to say, an emptiness so empty that its value is always calculated on the lowest tier of the spectrum of use; the dark margin, butt-end, and trash-speckled yard of a country on the move with no time to watch the clouds make shadows on the sand.

The political climate here also seemed attractive. Nevada has been from its inception, in the judgment of historian Gilman M. Ostrander, "the great rotten borough" of America—by which he meant a district whose withered population is far outstripped by its representation in the national legislature, and whose sovereignty thus gives it unique leverage. The state was added to the Union to give Abraham Lincoln an edge in the 1864 electoral college, as well as give extra Republican heft to Congress, and the gold mines outside Virginia City that George Hearst helped discover ensured a steady stream of hard currency to bring in fortune seekers and immigrants. By the time of the Great Depression, as instant towns like Rhyolite had dried up, the state was eager to find a new economic base and found itself born again as a legal cradle of all the vices that neighboring states did not officially want: namely gambling, prostitution, and liquor. It thereby defined itself as the dark corner of the country's psyche, a place for social and physical explorations into terra incognita; in the words of the essayist David Thomsen: "on the edge, on the wire, off to one side, in the empty quarter." The jerkwater highway stop of Las Vegas became a colored rebus for the American libido, the front yard of the Nevada Test Site where over one thousand nuclear weapons were detonated and federal jobs were welcomed by the local rookery of boosters.

But Yucca Mountain was a different story. This project smelled like a helping of poisoned pork, even though, at nearly $100 billion, it would have been the singlemost expensive construction effort in world history. One rule of the latter-day atomic age is that no politician who wants to hold on to power will let himself be perceived as anything less than a full-throated opponent to nuclear waste in the district. And so the pushback was fierce. The acquiescent time when Las Vegas welcomed the testing of hydrogen bombs on its northern horizon were over; what people remembered were the government's lies about fallout, as well as the dead sheep—and people—downwind. The Yucca Mountain plan, formalized in the 1987 Nuclear Waste Policy Amendment Act, became known to newspaper cartoonists as the "Screw Nevada" bill. Jawboning against it became an entry-level requirement for anyone trying to get elected in Nevada, second only to protecting casinos from more regulation. The governor set up a special office to fight the plan and to argue that groundwater seeping through the volcanic cracks inside the ridge would eventually corrode and crack the metal of the waste canisters, allowing toxic uranium, plutonium, neptunium, and a buffet of other elements to leak into the aquifer, ruining it forever.

The state of Nevada went so far as to quietly pay the attorney's fees for a Utah-based Indian tribe which was one of the only political bodies anywhere who actually wanted to be the host for nuclear garbage. The Skull Valley Band of Goshute occupy a small reservation on the edge of an alkaline plain south of the Great Salt Lake. Passing near Skull Valley by stagecoach on his way to the Nevada gold fields in 1867, Mark Twain called it "a vast waveless ocean stricken dead and turned to ashes." And in a section of his memoir *Roughing It* that has become legendary for its meanness, even for the bilious Twain, he made fun of the Goshutes for their poverty and their traditional diet of berries and insects. "Hungry, always hungry, and yet never refusing anything a hog would eat, though often eating what a hog would decline; hunters, but having no higher ambition than to kill and eat jackass rabbits, crickets and grasshoppers and to embezzle carrion from the buzzards and coyotes." The new century brought no prosperity. The Skull Valley Band found itself encircled by private waste dumps and a U.S. Army dump for nerve gas, and no jobs were around that didn't involve toxicity. The tribe began serious negotiations in 1996 with a group of electrical utilities to host a "monitored retrievable storage" facility for nuclear fuel—meaning the rods could be fished out and taken elsewhere at a later date, as opposed to being sealed in a sarcophagus like Yucca Mountain forever.

This deal would have been legal because, like other Indian tribes, the Skull Valley Band enjoys a limited degree of sovereignty: the same hazy juridical doctrine that had allowed tribes to open casinos on their reservations in the 1990s, even against the wishes of the states that encompass them. Each one of the 125 members of the tribe was to receive a reported $2 million payment in exchange for their support of nuclear waste on their homeland, which happened to be an hour's drive away from downtown Salt Lake City. The Utah legislature and the Mormon Church (often seen as coefficients, with some justification) reacted to this idea much as Nevada did to Yucca Mountain. What followed was heavy negative lobbying of every federal entity with a hand in the decision, from the Nuclear Regulatory Commission to the Bureau of Indian Affairs to the U.S. Air Force, which routinely flew F-16 fighter jets over the Great Basin on training missions. What if one of those jets should happen to fall from the sky and crash into an aboveground waste canister? At the urging of Utah officials, this one-in-a-million shot was duly studied for years and adjudged to be highly unlikely. But what finally killed the Skull Valley waste plan was the refusal of the Department of Interior in 2006 to approve the lease, partly because it also refused to grant right-of-way for trucks to carry the waste across government land.

Less than three years later, Yucca Mountain also appeared to be dead in the water. With little fanfare or explanation, newly elected President Barack Obama took away all the development funding for the project, leaving a skeleton staff in the Las Vegas offices and a vacant exploratory tunnel in the mountain where even the lights and the cooling have been turned off to conserve funds. The influence of Senate Majority Leader Harry Reid (D-Nevada) was partly credited for the reversal; indeed, on the night of the president's election, he jubilantly told reporters that the dump was "history." Energy Secretary Stephen Chu promised a congressional subcommittee that a panel of experts would start looking for new sites and new solutions. "We got a mighty expensive dinosaur out there," Representative Marion Berry, (D-Arkansas), lamented. The project has thus far cost $9 billion without a single gram of uranium finding its tomb. Almost the entire supply of the nation's nuclear waste still sits in a series of shallow pools and dry casks, many of which are immediately outside reactors and need constant maintenance.

Abe Van Luik was one of the lucky ones to keep his job at Yucca Mountain, and on the day I visited him, the office was mouse quiet. The accumulated research on "markers and monuments" will survive, though, and be passed along to whichever new team is assigned in the future to tackle the multigenerational problem left in the wake of the country's nuclear ambitions. This work had been pioneered in the mid-1990s by a group of experts advising the Waste Isolation Pilot Plant in Carlsbad, New Mexico, which included linguists, anthropologists, physicists, and one science fiction novelist—Gregory Benford, author of, among other titles, *The Jupiter Project*.

Part of their final recommendation was that the physical shape of the monument atop the nuclear waste be executed in a way to evoke "horror and sickness" and, in part, communicate the following:

> *Sending this message was important to us*
> *We considered ourselves a powerful culture*
> *This place is not a place of honor*
> *No highly esteemed deed is commemorated here*
> *Nothing valued is here*
> *What is here was dangerous and repulsive to us*

In order to be useful, the monument would have to outlast the collective memory of everyone on earth, as well as every natural and unnatural catastrophe to visit the planet in the next geologic era: global warming, volcanoes, continental drift, widespread social dissolution, the end of nation-states, the possible coming of alien spacecraft, the evolution of an entirely new sentient being for whom the time of humans will be like the time of pterodactyls is to us.

The entity called the "United States" has been around for less than one-fiftieth of the minimum decay time of the nuclear waste it plans to store at Yucca, yet built into some of the models is the assumption that Congress will exist in its current state and will be able to provide continuing funding to the Department of Energy (an admirable example of patriotism, perhaps, or just an agency that really can't bear to part with its budget).

What else besides a stone marker could survive the half-life of plutonium? Van Luik mused on the types of social structures that have endured in a communicative fashion across even the brief span of history since the time of Jesus Christ (one tenth the half-life), and the only answer is the Roman Catholic Church, which has gone through permutations across the decades but has retained the same basic message. Indeed, the panel in New Mexico has suggested employing a professional staff that will hand down knowledge of the malignant hell of nuclear waste from generation to generation, a kind of nuclear priesthood. "I guess if humanity stays around," says Van Luik, "they would be kind of like a religion."

He is more in favor of inert monuments—something on the order of the Pyramid at Cheops or the rock art on the cave walls at Laussane. Symbols that would demonstrate that those who erected the markers were intelligent beings with a knowledge of complex science. A few who have studied the issue have suggested detailed atomic diagrams of uranium and plutonium that would be recognizable to any civilization which has also tampered with the atom. Languages may be lost, but the code inside the building blocks of the universe remains constant—a kind of language in itself.

Such a plan would be reminiscent of the aluminum plaque fastened to the side of the 1972 Pioneer space probe sent toward the Taurus constellation. It featured a depiction of a naked man and woman, a diagram of the hydrogen atom to show we were clever, a map of the solar system and the position of our sun in relation to fourteen different pulsars across the galaxy—a map of how to find us. Such a transspecies greeting was designed to be a form of communication with whatever alien species might happen to intercept Pioneer in the intervening millions of years, and such is the spirit behind our future monument to nuclear waste.

The current plan at Yucca calls for carved warnings in seven different languages: English, Arabic, French, Spanish, Russian, Chinese, and Navajo. The assumption would be that even if these languages are lost to the collective memory that a smart intruder would be able to crack the code based on the patterns of letters and words—essentially a job of cryptography. But Van Luik has his doubts about whether this would work. Recalling the twelfth-century rock art on the canyon walls around Moab, Utah, which depicts bulbous stick figures of men amid squiggly lines and gazelles, he pointed out that the shapes are distinct, but the exact meaning has died along with the artists. "We can read them," he said, "but not *understand* them. What the actual messages are have been lost. The mindset is not the same."

In other words, our highly-evolved written language of today could be the nonsensical cave etchings of tomorrow. The long-range planning for nuclear waste now puts modern man in the conscious role of thinking of himself as an ancient civilization. Uranium has forced us to think of ourselves not as the vanguard of the species, but as "the old ones" whose purposes must be the subject of scholarship and archaeology. The time camera has been pulled back and the view is dizzying.

This is why Van Luik says he favors a set of granite temples that projects a nonverbal aura of something ominous, if not quite holy. "This would be a sad place," he said.

I asked him about the premise of the whole project. Does the United States feel an obligation to make these kinds of prognostications about societies in the distant future and warn them about the deadly legacy of what we left behind? Does this speak to nuclear guilt?

"Even if nature gives us a new species, you still don't want to do anything that damages them," he said. "We do not care about the purposeful intrusion," he added. "If this attracts people to the site and they get the message, that's fine. They will not be inadvertent intruders. What we are concerned about is the inadvertent intrusion"—that is, a group of futuristic miners or farmers or whoever who innocently sink a drill rig or a water well into a room full of deadly plutonium without being aware of the fatal treasure beneath.

But would future visitors to the ridge now known as Yucca Mountain be moved to heed these nuclearized monuments? Van Luik thinks this is a certainty. Anything dating from antiquity is bound to be a focus of curiosity—this, too, is a durable feature of humanity.

"If those monuments are ten thousand years old, they will have an aura attached to them," Van Luik said. "It will be like going to the Parthenon. You will have tours."

Not every suggestion for a monument makes sense. Dozens of artists have submitted their own proposals for timeless semiotics in the desert, with varying degrees of plausibility. One was of a giant toilet bowl, signifying the objectionable quality of what lies beneath. Another proposal from a professor at the University of Nevada at Las Vegas was made in all earnestness: a giant penis, which he said was an epochal symbol for territorialism and belligerence. Another monument involved a temple surrounded by jagged spikes coming from the earth at disquieting angles that suggest the insanity of what lies below. The New Mexico panel proposed a variety of exterior designs to mark this neo-Stonehenge. They were called, respectively: "Landscape of Thorns," "Spike Field," "Spikes Bursting Through Grid," "Leaning Stone Spikes," "Menacing Earthworks," "Black Hole," "Rubble Landscape," and "Forbidding Blocks."

Van Luik showed me his favorite wordless hieroglyph, a proposal that had come from a student at the Savannah College of Art and Design. A three-sided nuclear fan with a dark ray of energy comes from underground. Off to one side, a human figure bending to one knee in a pose that looks at first like reverence, but on closer inspection appears to be sickness and agony. Its effectiveness as a sign would, of course, be dependent on the cognizance of pain. Archaeologists have noted that the human form, even rendered as a stick figure, has always been one of the most durable artistic icons since the emergence of the species.

Several members of the New Mexico group had lobbied instead for a total obliteration of the site after it had been filled, counting on the sophisticated technology of tomorrow to be able to detect the waves of deadly energy coming from the earth. Building a monument on the site might have the opposite effect—it could inspire curiosity rather than repel it.

I told Van Luik that I could understand this argument. Creating a temple on the site might convince a visitor that this was a place of worship or burial of the ancient ones and inspire them to go exploring, just as Western archaeologists of the 1920s could not resist breaking into each and every pyramid left by the pharaohs of the Middle Kingdoms.

He acknowledged this was true. "We have dug like crazy to see what they were hiding," he said. "Every tomb we have raided. That's basic greed and curiosity."

Seven decades after the discovery of fission, there is much that remains enigmatic about uranium. Even a basic question—"How much of it does the U.S. government have under lock and key?"—does not have a firm answer.

The U.S. Department of Energy decided in 1996 to make a complete inventory of all the highly-enriched uranium America had manufactured through the years of the Cold War. Such a task may have seemed easy at first glance: the enrichment had taken place under heavy security and with strict accounting procedures. But it turned out that centralized records did not exist. The policy of "atomic secrecy" had created pockets of information exclusive to different divisions within the AEC and the Strategic Air Command and the U.S. Navy and several other civilian and military entities. Nobody ever had access to the total picture. A fragmentary portrait had to be assembled from logbooks and typed memoranda from each division in the uranium empire—Paducah, Oak Ridge, Hanford, Savannah River, and Rocky Flats—which began with the acquisition of the first barrels from Shinkolobwe.

The final report was suppressed for nearly a decade. The Federation of American Scientists waged a lengthy campaign to secure its release and when it was finally declassified, it showed that the United States had birthed slightly more than 1,000 tons of highly-enriched uranium in the course of half a century—and that approximately 3.2 tons of the stuff had vanished at some point.

The report's authors were careful not to blame the "inventory difference" on theft, explaining that measuring equipment was considered imprecise before the early 1970s and tiny differences between the book inventory and the actual product may have mounted over time. There was also what engineers call holdup—fragments of uranium gas that had clung to ducts and pipes and which can throw off the balance sheet. But there could be no definitive explanation for the loss.

The disparity was even greater on the other side of the cold-war divide. After the collapse of the Soviet Union, American researchers were surprised to learn that the Kremlin had *no idea* how much of the material it had produced during its four-decade buildup. David Albright, president of the Institute for Science and International Security and a former weapons inspector himself, put the total Russian stockpile at anywhere between 735 and 1,365 tons. This estimate was made during Boris Yeltsin's presidency a decade ago, the last time when anything resembling transparency existed in Russia. The gap of more than 600 tons represented a supply of uranium as heavy as an ocean frigate and enough to make more than eight thousand nuclear weapons of the size and type that leveled Hiroshima. American investigators had no idea how much of it still existed, or where it might be stored.

"We're not sure even *they* understand how much they have," Albright said of the Russians. Their inexactitude, he said, was due to poor record keeping over the decades.

His thoughts were echoed by Laura Holgate of the Nuclear Threat Initiative, a Washington, D.C.–based organization dedicated to warning governments and the public about the dangers of loose nuclear goods. An accurate tally of the amount of highly enriched uranium produced by Russia is impossible, thanks to poor accounting and a culture of petty mistruth during the days of Communism. Rigorous quotas had encouraged plant managers to hoard a little uranium in case of a seasonal slowdown and to avoid criticism from their superiors in Moscow. The official logbooks were routinely falsified and the production estimates therefore have a huge margin of error.

"There truly is no knowing how much really exists," she said.

Holgate is a pleasant woman in her early forties with librarian's glasses and an easy laugh. She told me one of the most frightening moments of her career was being taken inside a vault at the Oak Ridge National Laboratory in Tennessee. Inside were rows of multiple shelves—

like a library—and resting in metal cradles on these shelves were the secondaries for hydrogen bombs. These were the charges, coated with raw uranium and cored with enriched uranium that provided an explosion of 50 million degrees Fahrenheit to force hydrogen nuclei to fuse together, creating a fireball hotter than the interior of the sun.

To Holgate, the shelves seemed to go on and on.

The uranium archipelago that France operates on the African continent used to include several mines in its former colony of Gabon, a small dictatorship at the edge of the South Atlantic. The uranium deposits are deep in the interior, and one of the more productive mines was at a spot near the equator called Oklo.

A chemist in France was examined some samples from the mine in 1972 and noticed that the proportion of U-235 was slightly lower than the usual concentration of 0.7 percent. This figure was thought to be a constant throughout the world, a rate set by the unchanging half-life of the uranium atom, and so the findings were highly unusual. The company informed the French Commissariat à l'Énergie Atomique, which started its own probe into the matter. Their conclusions were startling.

The soil at Oklo had been a natural nuclear reactor. Approximately two billion years ago, the proportion of U-235 on the earth had been as high as 3 percent. The deposits at Oklo were sandwiched between layers of sandstone and granite and tilted in such a way that water could flow through the cracks and create pockets of highly concentrated ore. The water also acted as a moderator, slowing down the flying neutrons just enough to make them hit the uranium nuclei and create an underground chain reaction. The heat made the water turn to steam, which calmed the reaction, and when the water condensed again, the reaction restarted. This may have gone on for a million years before the uranium depleted, and accounted for the low number of U-235 atoms in the sample sent to France. They had simply been destroyed two billion years ago in the Precambrian era.

At roughly the same time, simple bacteria in the nearby oceans were learning to respire, using oxygen to convert food into energy for the first time. The underground chain reaction ran concurrent with the rise of life on earth.

"After the reactor had shut down, the evidence of its activity was preserved virtually undisturbed through the succeeding ages of geological activity," wrote George Cowan, a former Manhattan Project scientist who visited the site.

He concluded: "In the design of fission reactors, man was not an innovator but an unwitting imitator of nature."

The vegetable oil warehouse that sheltered Edgar Sengier's barrels was demolished years ago, but the lot where it stood can still be found on the shores of Staten Island. The property is located in a scruffy waterfront neighborhood at the southern foot of the Bayonne Bridge, dotted with derelict lots and auto supply companies. This had once been a busy part of New York Harbor before the rise of large container ships had consigned it to obscurity.

I had been given the address and went to go see it one night. There was a light fog, muting the lights on the span of the crescent-shaped bridge. Slightly to the east was a vacant yard, a postindustrial gloaming empty of everything but scrap metal, a few parked vehicles, and a stack of concrete road barriers. There was a fence crowned with barbed wire and some roadside bushes concealing a placard: PROTECTED BY SUPREME SECURITY SERVICES: TOTAL SECURITY FOR AN INSECURE WORLD. 877-877-7899. Another sign warned of a guard dog protecting the emptiness.

The northern edge of the property fronts a tidal strait called Kill Van Kull, which separates Staten Island from Bayonne, New Jersey. Most of the remnants of the World Trade Center had been towed by barge through this channel in the autumn of 2001, on their way to permanent disposal in a nearby landfill. The debris had passed only a few yards away from the lot where the material for the first atomic bomb had lain in waiting a little less than sixty years before.

I stood at the edge of the lot and tried to imagine what the place might have looked like when the barrels full of yellow African dirt had been unloaded from their freighters and rolled into the warehouse. This was only the second continent it would touch. The uranium was in for two years of senescence, a brief sleep, before it would be trucked south for enrichment and then dropped over an Asian city, changing the world as it fell.

ACKNOWLEDGMENTS

Hundreds of people were generous with their time and expertise in the research and reporting of this book. Thanks in particular to Nadezda Kavalirova of the Confederation of Political Prisoners in Prague, Jiri Pihera of the Tüv Cert mining company in St. Joachimsthal, Herman Meinel of the Museum Uranbergau in Schlema, Amanda Buckley of Rio Tinto in Darwin, and Moussa Abdoulaye of Areva in Niamey. Steve Kidd of the World Nuclear Institute in London shared his considerable knowledge, gave spirited and good-natured banter, and pointed out a handful of mistakes in the hardcover version. Jennifer Steil of the *Yemen Observer* was a gracious host in Sana'a, and I am extremely grateful for her referrals and suggestions. Irina Lashkhi of the Open Society Institute pried open some government doors in Tbilisi.

Dr. Sung Kyu Kim of Macalester College and Dr. Keith Olive of the University of Minnesota helped me understand some points of atomic physics. Robert Alvarez offered some good advice. Mark Steen spent many hours on the telephone with me, telling stories about his famous father and offering wisdom about the current state of the uranium business. Vilma Hunt and Robert S. Norris—first-class uraniumophiles, both—provided inspiration and encouragement. Namposya Nampanya-Serpell supplied the Bemba definition of the word *Shinkolobwe*. David Schairer passed along the curious history of the Maria Theresa thaler.

David Smith of the New York Public Library is a magician among librarians, who can pull obscurities from that mammoth collection better than anybody I know. Walt and Sandy Lombardo, who hosted a reading for me in their Las Vegas bookstore in the summer of 2006, have been good friends, as well as a valuable source of geologic knowledge. Rainer Karlsch, an expert on Wismut history, offered advice in Berlin.

I am also thankful to Greg Cullison, Danielle Dahlstrom, Laura Babcock, Pamela Carr, Luke Newton, Alexia Brue, Nadine Rubin, Meeghan Truelove, Russ Baker, Andrew Bast, Nina Nowak, Sugi Ganeshananthan, Tom Vanderbilt, Alexis Washam, Michael Hawkins, Lionel Martin, Tungalag Flora, Sambalaibat, Erdenejargal Perenlei, Terry Wetz, Ron Hochstein, Doug Kentish, Martin Eady, Robert Holland, Tom Pool, Curt Steel, John Cassara, Monte Paulson, Bob Etter, Terry Babcock-Lumish, Gabrielle Giffords, Bill Carter, Julie McCarthy, Beverly Bell, Rinku Sen, Marybeth Holleman, Jim K. Cambon, Kevin D'Souza, Malcolm Shannon, Johanna Lafferty, Leslie Najarian, Frederic D. Schwarz, Iftikhar Dadi, William Finnegan, and Tara Parker.

Dr. Charles Blatchley of the physics department at Pittsburg State University read a late draft of the manuscript and offered valuable suggestions, as did George White, formerly of Nuexco Information Services, and Robert S. Norris of the Natural Resources Defense Council. The improvements belong to them, and I claim every error.

A substantial portion of this book was written at the Mesa Refuge writers' colony overlooking Tomales Bay in California. I am grateful to the Common Counsel Foundation for its support.

My family in Arizona believed in me.

Martha Brantley provided exceptional encouragement during the writing of this book and I will always be grateful to her. Kevin Gass traveled with me to Yemen and Georgia, suffered through my uranium stories, made me laugh when I needed it, and spurred me on through the rough patches. It would be difficult to conceive of a finer man for a best friend.

Kathryn Court of Viking Penguin was the editor of this book, and that has meant everything. She is a smart reader, a careful editor, and a delightful person in general. Branda Maholtz of Viking Penguin also gave the manuscript her thoughtful scrutiny and made it sharper. Deborah Weiss Geline made many wise suggestions during copyediting, and Bruce Giffords made it come together in production. Literary agent Brettne Bloom made it happen in the first place.

New York City–Point Reyes Station, California–
Hanover, New Hampshire
January 2007–July 2009

NOTES ON SOURCES

Translators were employed for some of the interviews conducted overseas, and the contents of a few of those conversations have been compressed into single running paragraphs with the back-and-forth questioning eliminated for the sake of readability.

A library the size of a small city might be necessary to house all the books, magazine articles, newspaper stories, academic papers, technical manuals, films, and government reports that have documented how uranium has moved the world in various ways. I used a tiny fraction of that available material in this book, and what follows is a guide to the sources from which I drew statistics, quotes, details, and anecdotes. Complete endnotes and citations can be found at www.tomzoellner.com.

INTRODUCTION AND CHAPTER 1: SCALDING FRUIT

The detail about the Japanese company at Temple Mountain is from "From X-Rays to Fission, A Metamorphosis in Mining," by Clay T. Smith, in *Geology of the Paradox Basin Fold and Fault Belt, Third Field Conference* (Durango, Colo.: Four Corners Geological Society, 1960). The image of twitching sand is from the physicist Otto Frisch and was taken from *The Making of the Atomic Bomb*, by Richard Rhodes (New York: Simon & Schuster, 1986). Forced-labor policies in the Congo were discussed in "Pouch Letter 39," a declassified memo written by the American OSS agent Wilbur O. Hogue, sent on July 5, 1944, and on file in the National Archives, "Records of the Office of Strategic Services," Record Group 226, Entry 108C. Living conditions at Union Minière mine properties were disclosed in *L'histoire du Congo, 1910–1945*, by Jules Marchal (Borgloon, Belgium: Editions Paula Bellings, 1999), and in *The Creation of Elisabethville, 1910–1940*, by Bruce Fetter (Stanford, Calif.: Hoover Institution, 1976). Other details are in the official company history *La Mangeuse de Cuivre: La Saga de l'Union Minière du Haut-Katanga, 1906–1966*, by Fernand Lekime (Brussels: Didier Hatier, 1992). I am grateful to Pieterjan Van Wyngene for locating and translating the relevant sections of these books for me. Historical background on Shinkolobwe is in "Rip Veil from Belgian Congo Uranium Mine," a Reuters dispatch reprinted in the *Chicago Daily Tribune*, Nov. 7, 1956; "Shinkolobwe: Key to the Congo," by Ritchie Calder, in the *Nation*, Feb. 25, 1961; and "Africa Holds Key to Atomic Future," by George Padmore, in the *Chicago Defender*, Sept. 8, 1945. More information on Shinkolobwe, including the *Bloc Radioactif!* anecdote, came from *Inside Africa*, by John Gunther (New York: Harper & Row, 1953). A vivid fictional account of Elisabethville is in the novel *Radium*, by Rudolf Brunngraber (London: G. G. Harrap, 1937). Larry Devlin's memories are taken from a personal conversation with the author, as well as from his memoirs, *Chief of Station, Congo: Fighting the Cold War in a Hot Zone* (New York: Public Affairs, 2007). Some technical details of the recent activity at Shinkolobwe were taken from the report *Assessment Mission of the Shinkolobwe Uranium Mine, Democratic Republic of Congo*, written by the Joint United Nations Environment Program and the UN Office for the Coordination of Humanitarian Affairs, dated 2004 and published in Geneva. A look at the underside of Congolese mineral trading is found in the July 2006 report *Digging in Corruption: Fraud, Abuse, and Exploitation in Katanga's Copper and Cobalt Mines*, researched and published by Global Witness in London, as well as *The State vs. The People: Governance, Mining, and the Transitional Regime in the Democratic Republic of Congo*, by Netherlands Institute for Southern Africa, and published in Amsterdam in 2006. Illicit uranium trading is probed in "Letter Dated 18 July 2006 from the Chairman of the Security Council Committee Established Pursuant to Resolution 1533 (2004) Concerning the Democratic Republic of the Congo Addressed to the President of the Security Council, Conveying the Report of the Group of Experts," available from the United Nations in New York. The issue is further explored in the Congolese government intelligence brief "Unofficial Exploitation of Uranium at the Shinkolobwe Mine in Katanga" [date and author unknown], and the news stories "'Uranium' Seized in Tanzania," from the British Broadcasting Company, Nov. 14, 2002, and "Iran's Plot to Mine Uranium in Africa," in London's *Sunday Times*, Aug. 6, 2006. Details on the Kinshasa reactor are from "Missing Keys, Holes in Fence, and a Single Padlock: Welcome to Congo's Nuclear Plant," by Chris McGreal, in the *Guardian*, Nov. 23, 2006. General information on Mobutu's rule and the exploitation of Katanga is in Martin Meredith's *The Fate of Africa: A History of Fifty Years of Independence*

(New York: Public Affairs, 2005). The practice and origins of the term *Article Fifteen*, information on the sorry state of the Kinshasa reactor, and other fascinating details on Mobutu's reign and downfall are drawn from Michela Wrong's *In the Footsteps of Mr. Kurtz: Living on the Brink of Disaster in Mobutu's Congo* (New York: HarperCollins, 2001). An unforgettable account of life and death in the colonial period is in *King Leopold's Ghost*, by Adam Hochschild (New York: Mariner Books, 1998).

CHAPTER 2: BEGINNINGS

This chapter and the next benefited greatly from Richard Rhodes's *The Making of the Atomic Bomb* (New York: Simon & Schuster, 1986). Multiple facts and anecdotes—including Ernest Rutherford's experiments, Otto Frisch's walk through the Swedish woods, and the comparison of the Manhattan Project to the auto industry—were taken from Rhodes's text-dense volume, easily the best comprehensive account of this often-told chapter in world history. Some history of St. Joachimsthal is in the town history *Jachymov: The City of Silver, Radium, and Therapeutic Water*, by Hana Hornatova (Prague: Medeia Bohemia, 2000); in the article "The Silver Miners of the Erzgebirge and the Peasants' War of 1525 in Light of Recent Research," by George Waring, in *Sixteenth Century Journal*, Summer 1987; and in "History of Uranium," by Fathi Habashi and Vladimir Dufek, in the *CIM Bulletin*, Jan. 2001, published by the Canadian Institute of Mining, Metallurgy, and Petroleum. Additional details on the St. Joachimsthal Valley were taken from *Atomic Rivals: A Candid Memoir of Rivalries Among the Allies Over the Bomb*, by Bertrand Goldschmidt (New Brunswick, N.J.: Rutgers University Press, 1990). The peasants' song was in *Uranium Matters: Central European Uranium in International Politics, 1900–1960*, by Zbynek Zeman and Rainer Karlsch (Budapest: Central European University Press, 2008). The Taborite movement is explored in *Pursuit of the Millennium*, by Norman Cohn (Fairlawn, N.J.: Essential Books, 1957). Martin Klaproth's story and Otto Frisch's encounter with finished blocks of U-235 in chapter 3 were drawn from *The Deadly Element*, by Lennard Bickel (Hong Kong: Macmillan, 1979). Early musings about atomic power are found in the disjointed but prescient novel *The World Set Free: A Story of Mankind*, by H. G. Wells (Leipzig: Bernhard Tauchnitz, 1914), and the farsighted treatise *The Interpretation of Radium*, by Frederick Soddy (London: John Murray, 1912). Radium physics are discussed in the articles "Radium and Radioactivity," by Marie Curie, in *Century* magazine, Jan. 1904, and "Radiation Hormesis," by Jennifer L. Prekeges, in the *Journal of Nuclear Medicine Tech-*

nology, Nov. 2003. Some atomic metaphors were inspired by and drawn from *A Short History of Nearly Everything,* by Bill Bryson (New York: Broadway Books, 2004), and some physics were drawn from *Atom: Journey Across the Subatomic Cosmos,* by Isaac Asimov and D. H. Bach (New York: Plume, 1992). Brief biographies of the Curies, Ernest Rutherford, Henri Becquerel, and James Chadwick and copies of their Stockholm lectures are archived at http://nobelprize .org. Background on the first man to use the phrase *atomic bomb* is from *H. G. Wells: His Turbulent Life and Times,* by Lovat Dickson (Middlesex, UK: Penguin, 1969); *The Science Fiction of H. G. Wells,* by Frank McConnell (Oxford: Oxford University Press, 1981); and *H. G. Wells,* by Norman and Jeanne MacKenzie (New York: Simon & Schuster, 1973). Early assessments of uranium are in "Vast Energy Freed by Uranium Atom," the *New York Times,* Jan. 31, 1939, and "The Atom Is Giving Up Its Mighty Secrets," by Waldemar Kaempffert, the *New York Times,* May 8, 1932. Otto Hahn's despair was mentioned in *Savage Dreams,* by Rebecca Solnit (New York: Vintage, 1994). Ed Creutz's quote came from *Atomic Quest,* by Arthur Compton (Oxford: Oxford University Press, 1956).

CHAPTER 3: THE BARGAIN

The recollections of Robert Rich Sharp are preserved in his memoirs, *Early Days in Katanga* (Bulawayo, Southern Rhodesia: Rhodesia Printers, 1956). Leslie Groves discloses a portion of what he knew, including his relationship with Edgar Sengier, in his *Now It Can Be Told* (New York: Harper & Brothers, 1962). Robert Laxalt's recollection of the OSS officer is found in his memoirs, *A Private War: An American Code Staffer in the Belgian Congo* (Reno: University of Nevada Press, 1998). Portions of the Guarin geological report are excerpted in the anthology *The Secret History of the Atomic Bomb* (New York: Dell, 1977). The lawyer-adverse quote from Edgar Sengier comes from *Uranium Trail East,* by Cordell Richardson (London: Bachman & Turner, 1977). An excellent map and description of the Archer Daniels Midland facility on Staten Island can be found in the CD book *The Traveler's Guide to Nuclear Weapons Sites,* by Timothy L. Karpin and James M. Maroncelli (Lacey, Wash.: Historical Odysseys Publishers, 2002). The complaint about secrecy is drawn from a memo from Joe Volpe to Phillip Merritt, dated Jan. 3, 1945, and on file in the National Archives, "Manhattan Engineering District, General Administrative Files, General Correspondence 1942–1948," Record Group 77, Entry 5. The OSS assessment of the Congo is in a memo dated Oct. 31, 1944, and on file in the National Archives, "Records of the Office of Strategic Services," Record Group 226,

Entry 108C. The story of the Leipzig fizzle was taken from *The Making of the Atomic Age,* by Alwyn McKay (Oxford: Oxford University Press, 1984). Some aspects of wartime enrichment at Oak Ridge and Hanford were taken from "The First Fifty Years," in the *Oak Ridge National Laboratories Review* 25, no. 3–4 (1992). Details about Edgar Sengier and the operation at Shinkolobwe, including the financing of the pit rehabilitation, are drawn from *Gathering Rare Ores: The Diplomacy of Uranium Acquisition, 1943–1954,* by Jonathan E. Helmreich (Princeton, N.J.: Princeton University Press, 1986). The conversation between Sengier and Colonel Kenneth Nichols has been reconstructed based on two sources: Nichols's memoirs, *The Road to Trinity,* by Kenneth D. Nichols (New York: Morrow, 1987), and the official company history, *La Mangeuse de Cuivre: La Saga de l'Union Minière du Haut-Katanga, 1906–1966,* by Fernand Lekime (Brussels: Didier Hatier, 1992). The story of the mine at Port Radium comes from *Flame of Power: Intimate Profiles of Canada's Greatest Businessmen,* by Peter Charles Newman (Toronto: Longmans, Green and Company, 1959), and the superb self-published autobiography of mining engineer Fred Chester Bond. This memoir is called *It Happened to Me,* and was provided to me by Bruce Bond, the son of the author. U.S. Navy documents from the National Archives detailing the strange story of U-234 have been reproduced in the self-published book *Critical Mass: How Nazi Germany Surrendered Enriched Uranium for the United States Atomic Bomb,* by Carter Hydrick (Houston, Tex.: JiffyLine, 1998). Hydrick embraces some maverick theories, particularly that Germany possessed highly enriched uranium. His navy documents regarding U-234's voyage, however, can be regarded as beyond question. A rigorous analysis of the known facts is in *Germany's Last Mission to Japan: The Failed Voyage of U-234,* by Joseph Mark Scalia (Annapolis, Md.: Naval Institute Press, 2000). The quotes from John Lansdale and Hans Bethe were taken from the 2001 documentary film *Hitler's Last U-Boat,* directed by Andreas Gutzeit and distributed by International Historic Films, Inc., of Chicago. Further background and statements from Lansdale were taken from "Captured Cargo, Captivating Mystery," by William J. Broad, the *New York Times,* Dec. 31, 1995, and the obituary "John Lansdale, 91, Hunter of Nazi Atomic-Bomb Effort," by Anahad O'Connor, the *New York Times,* Sept. 3, 2003. Several details about Hanford and the life of Leslie Groves were taken from the excellent biography *Racing for the Bomb: Leslie R. Groves, the Manhattan Project's Indispensable Man,* by Robert S. Norris (Hanover, N.H.: Steerforth Press, 2003). Norris kindly furnished me with the daybook entry for Groves's Aug. 13, 1945, conversation with Edwards. His book was also the source for the story of the enterprising reporter (and would-be draftee) Thomas Raper of Cleveland. The gadget anecdote came from *109 East Palace: Robert Oppenheimer and the Secret City of Los Alamos,* by Jenant Conant (New York: Simon & Schuster, 2005). I also drew facts from *American Prometheus: The Triumph and Tragedy of J. Robert Oppenheimer,* by Kai Bird and Martin Sherwin (New York: Knopf, 2005); "Now They Can Be Told Aloud, Those Stories of 'The Hill,'" by William McNulty, in the *Santa Fe New Mexican,* Aug. 7, 1945; and *The Manhattan Project: Big Science and the Atom Bomb,* by Jeff Hughes (New York: Columbia University Press, 2002). The two-part Rabi quote is from "After the Bomb, A Mushroom Cloud of Metaphors," by James Gleick, in the *New York Times,* May 21, 1989; and also *Picturing the Bomb,* by Rachel Fermi and Esther Samra (New York: Harry N. Abrams, 1995). The recollections from the survivors of Hiroshima were drawn from Robert Jay Lifton's *Death in Life* (New York: Random House, 1967) and quoted in *The Making of the Atomic Bomb,* by Richard Rhodes (New York: Simon & Schuster, 1986). Further details were from the article "Imagining Nuclear Weapons: Hiroshima, Armageddon, and the Annihilation of the Ichijo School," by James Foard, in the *Journal of the American Academy of Religion,* 1997. The weight comparison between the Nagasaki plutonium and the penny comes from *The Curve of Binding Energy,* by John McPhee (New York: Farrar, Straus and Giroux, 1973). Of special note are the memoirs of Otto Frisch, modestly and inaccurately titled *What Little I Remember* (Cambridge: Cambridge University Press, 1980), from which several anecdotes in this section—including the stacked oranges in Richmond and the dragon-tickling experiment at Los Alamos—were taken. One of the unexpected pleasures I had in writing this book was reading Otto Frisch, who was as subtle and engaging as a writer as he was insightful as a theorist.

CHAPTER 4: APOCALYPSE

An early forecast for atomic power is in the Associated Press report "Fantastic World Envisioned with Atomic Energy," by Chiles Coleman, reprinted in, among other papers, the *Corpus Christi Caller-Times,* Aug. 7, 1945. Some editorial clips came from the peerless work *By the Bomb's Early Light* by Paul S. Boyer (Chapel Hill: University of North Carolina Press, 1994). Saint Augustine's admonition was recalled in "A Comet's Tale," by Tom Bissell, in *Harper's,* Feb. 2003. Some fragmentary pictures of the post-Hiroshima mood are in "U.S. Warns Artist Who Sketched Bomb," in the *Philadelphia Inquirer,* Aug. 8, 1945; "Wellsian Apocalypse," in the *Washington Post,* Oct. 25, 1945; "Atom Discounted as Rival of Coal," in the *Philadelphia Inquirer*, Aug. 8, 1945; and the Associated Press dispatch "Vatican City Paper Deplores Creation of 'Catastrophic' Weapon," Aug. 7, 1945. The Morley quote appeared in *Human Events* on August 29, 1945, and was republished in *The Manhattan Project*, ed. Cynthia C. Kelly (New York: Black Dog & Leventhal Publishers, 2007). The grim poll results from the Social Science Research Center were quoted in *Life Under a Cloud: American Anxiety About the Atom*, by Allan M. Winkler (Oxford: Oxford University Press, 1993). This excellent book also provided source material for this chapter, including Oppenheimer's first conversation with Harry S. Truman and a sketch of Bernard Brodie's thinking. Brodie's philosophy is also examined in *Bernard Brodie and the Foundations of American Nuclear Strategy*, by Barry Steiner (Lawrence: University Press of Kansas, 1991). Some of the speculations about the power of uranium are in "Uranium Metal," in *Scientific American*, Feb. 1947; "New Responsibilities," in *Science News Letter*, Aug. 18, 1945; "Congo Is Blind to the Richest Uranium Mine," by Associated Press reporter Arthur L. Gavshon, in the *Washington Post*, Aug. 6, 1950; and "Mystery Man of the A-Bomb," by John Gunther, in *Reader's Digest*, Dec. 1953. Reports of a uranium black market in China came from the Jan. 20, 1946, memo "Directive to All X-2 Field Stations" and the June 24, 1946, summary report from X-2's Shanghai station. Both reports are located in the National Archives in "Records of the Office of Strategic Services," Record Group 226, Entry 211. The account of Oppenheimer's ill-fated meeting with Truman is in Winkler, *Life Under a Cloud*, as well as *109 East Palace: Robert Oppenheimer and the Secret City of Los Alamos*, by Jenant Conant (New York: Simon & Schuster, 2005). The major writings of William L. Laurence are his two memoirs, *Dawn Over Zero* and *Men and Atoms*, as well as the *New York Times* stories "U.S. Atom Bomb Site Belies Tokyo Tales," Sept. 12, 1945, and "Atom Bombing of Nagasaki Told by Flight Member," Sept. 9, 1945. His article "The Atom Gives Up" was in the *Saturday Evening Post*, Sept. 7, 1940. He offers candid reflections in two oral history interviews, the transcripts of which are on file in the Rare Book and Manuscript Library at Columbia University in New York City. One interview was conducted by Scott Bruns in 1964; the other by Louis Starr in 1957. A bit more of his biography, including the anecdote about the rifle butt, is in his obituary "William Laurence of *The Times* Dies," on Mar. 19, 1977, as well as in an essay he wrote for the anthology *How I Got That Story*, by members of the Overseas Press Club (New York: E. P. Dutton, 1967). The influence of his hyperbole is examined in *The Myths of August*, by Stewart Udall (New York: Pantheon, 1994); *News Zero: The New York Times and the Bomb*, by Beverly Ann Deepe Keever (Monroe, Maine.:

Common Courage Press, 2004); a later article by Keever, "Top Secret: Censoring the First Rough Drafts of Atomic-Bomb History," in the journal *Media History*, vol. 4, no. 2, 2008; *Nuclear Fear: A History of Images*, by Spencer R. Weart (Cambridge, Mass.: Harvard University Press, 1988); and, most especially, in the peerless *Hiroshima in America: A Half Century of Denial*, by Robert Jay Lifton and Greg Mitchell (New York: Avon Books, 1995). His post-Hiroshima reporting on radioactivity was scrutinized in *The Exception to the Rulers: Exposing Oily Politicians, War Profiteers, and the Media That Love Them*, by David Goodman and Amy Goodman (New York: Hyperion, 2004); an excerpt has been republished at commondreams.org. The testing outside of Las Vegas is examined in Weart, *Nuclear Fear*; "The Mushroom Cloud as Kitsch," by A. Costandina Titus, in the anthology *Atomic Culture: How We Learned to Stop Worrying and Love the Bomb*, edited by Scott C. Zeman and Michael A. Amundson (Boulder: University Press of Colorado, 2004); *Bombs in the Backyard: Atomic Testing and American Politics*, by A. Costandina Titus (Reno: University of Nevada Press, 2001); and *Nukespeak: The Selling of Nuclear Technology in America*, by Stephen Hilgartner, Richard Bell, and Rory O'Conner (San Francisco: Sierra Club Books, 1982). Further period details of Las Vegas and the Nevada Test Site are in "The Melted Dog: Memories of an Atomic Childhood," by Judith Miller, in the *New York Times*, Mar. 30, 2005, and Vanderbilt, *Survival City*. The army quote about "oblivion" was from *Combat Zoning*, by David Loomis (Reno: University of Nevada Press, 1993), which was also quoted in *Blank Spots on the Map*, by Trevor Paglen (New York: Dutton, 2009). The General Electric executive's quote comes from "Power from the Atom: An Appraisal," by C. G. Suits, in *Nucleonics*, Feb. 1951, and was quoted in "Atomic Myths, Radioactive Realities: Why Nuclear Power Is the Poor Way to Meet Energy Needs," by Arjun Makhijani, in the *Journal of Land, Resources & Environmental Law* 24, no. 1 (2004). John C. Frémont's observations were quoted in *Savage Dreams*, by Rebecca Solnit (New York: Vintage, 1994). Changes to American architecture and planning were skillfully examined in *Survival City: Adventures Among the Ruins of Atomic America*, by Tom Vanderbilt (Princeton, N.J.: Princeton Architectural Press, 2002), which was also the source of the Augur quotes and the "terrible light" fragment from *Time*. E. B. White's passage is from his short and delightful book *Here Is New York* (New York: Harper & Brothers, 1948). Project Gnome is discussed in "Peaceful Atomic Blasting," in *Time*, Mar. 4, 1958; "Radiation Drops in A-Blast Zone," by Bill Becker, in the *New York Times*, Dec. 11, 1961; and "U.S. A-Bomb Test Releases Radiation," by Bill Becker, in the *New York Times*, Dec. 10, 1961; *Beyond Engineering: How*

Society Shapes Technology, by Robert Pool (Oxford: Oxford University Press, 1997); and in *Proving Grounds: Project Plowshare and the Unrealized Dream of Nuclear Earthmoving*, by Scott Kirsch (New Brunswick, N.J.: Rutgers University Press, 2005). Criticisms of the Atoms for Peace program are also noted in *The Curve of Binding Energy*, by John McPhee (New York: Farrar, Straus and Giroux, 1973). Mount Weather's layout is discussed in "Is This Bush's Secret Bunker?," by Tom Vanderbilt, in the *Guardian*, Aug. 26, 2006. Some cold-war reckonings were taken from *Atomic Audit: The Costs and Consequences of U.S. Nuclear Weapons Since 1940*, edited by Stephen I. Schwartz (Washington, D.C.: Brookings Institution Press, 1998); and *Arsenals of Folly: The Making of the Nuclear Arms Race*, by Richard Rhodes (New York: Knopf, 2007), which was also the source of the Churchill quote. Background on Abraham Feinberg is in "Going Steady," in *Time*, Aug. 29, 1955, as well as Michael Karpin's excellent *The Bomb in the Basement* (New York: Simon & Schuster, 2006), one of the most comprehensive accounts of the Israeli nuclear program's history. The scrambled eggs anecdote involving Ernst Bergmann, as well as other details of Dimona's origins and the description of the reprocessing building, were taken from *The Samson Option: Israel's Nuclear Arsenal and American Foreign Policy*, by Seymour M. Hersh (New York: Random House, 1991). Further background was drawn from the June 1981 United Nations report "Israel: Nuclear Armament," prepared by Ali Mazuri et al.; the seminal *Israel and the Bomb*, by Avner Cohen (New York: Columbia University Press, 1998); "Recipe for an Israeli Nuclear Arsenal," by Martha Wegner, in *Middle East Report*, Nov. 1986; and the article "The Nuclear Arsenal in the Middle East," by Frank Barnaby, in *Journal of Palestine Studies*, Autumn 1987, which was the source of the "pushed into the sea" observation. Mordechai Vanunu's story was first told in "Revealed: The Secrets of Israel's Nuclear Arsenal," in the London *Sunday Times*, Oct. 5, 1986, and the tale of "Cindy" and her Roman flat is related in "Mordechai Vanunu," the *Guardian*, Apr. 16, 2004. The Hounam quote comes from his book *The Woman from Mossad: The Story of Mordechai Vanaunu and the Israeli Nuclear Program* (Berkeley, Calif: Frog Books, 2000). The *Scheersberg* story first broke in the *Los Angeles Times*, "200 Tons of Uranium Lost; Israel May Have It," by Robert Gillette, Apr. 29, 1977, followed up in the *New York Times* with "Escort Unit Urged for Uranium Cargo," Apr. 30, 1977; a *New York Times* op-ed titled "The Plumbat Affair," by Paul S. Leventhal, Apr. 30, 1978; and also "Uranium Loss Fails to Change Security," in the *New York Times*, Feb. 25, 1979. *Time* magazine exposed more details in "Uranium: The Israeli Connection,"

May 30, 1977. The story of A. Q. Khan has been told in several places, but the following sources contributed most heavily to this section: the 2007 dossier *Nuclear Black Markets: Pakistan, A. Q. Khan, and the Rise of Proliferation Networks*, by the staff of the International Institute for Strategic Studies in London; the book *The Atomic Bazaar*, by William Langewiesche (New York: Farrar, Straus and Giroux, 2007), from which I drew the color about Khan's relationship with Frits Veerman, among several other details; the British Broadcasting Company documentary "The Nuclear Wal-Mart," reported by Jane Corbin, aired on Nov. 12, 2006; "A Tale of Nuclear Proliferation," in the *New York Times*, Feb. 12, 2003; and *The Nuclear Jihadist*, by Douglas Frantz and Catherine Collins (New York: Twelve Books, 2007), which was the source for several facts, including "the Beast" in Johannesburg, Khan's largess to journalists and charities, and some snippets from his authorized biography, as well as the Bob Hope joke quoted elsewhere in this chapter. Some centrifuge metaphors were from "A Tantalizing Look at Iran's Nuclear Program," by William J. Broad, in the *New York Times*, Apr. 29, 2008. Some of Khan's stranger investments were disclosed in "Khan Built Hotel in Timbuktu," in the *Times of India*, Feb. 1, 2004. Khan's jingoistic quote was taken from *Pakistan 1995*, by Rashul Rais and Charles Kennedy (Denver: Westview Press, 1995). His recollection of free chai was in the Q&A "Abdul Qadeer Khan: The Man Behind the Myth," by Zeba Khan, reprinted by the Human Development Foundation at yespakistan.com. Some of the flavor of Dubai is captured in "An Unlikely Criminal Crossroads," in *U.S. News & World Report*, Dec. 5, 2005; "Boom Town," in the *Guardian*, Feb. 13, 2006; and the well-reported memoir *Hide and Seek: Intelligence, Law Enforcement, and the Stalled War on Terrorist Finance*, by John Cassara (Dulles, Va.: Potomac Books, 2006). Some post-Khan impacts were examined in "The Bomb Merchant," by William J. Broad and David Sanger, in the *New York Times*, Dec. 26, 2004, and "How Gadhafi Got His Groove Back," by Judith Miller, in the *Wall Street Journal*, May 16, 2006. Information on the uranium mine at Banjawarn comes from "A Case Study on the Aum Shinrikyo," by the staff of the U.S. Senate Government Affairs Permanent Subcommittee on Investigations, Oct. 31, 1995; *Aum Shinrikyo, Al-Qaeda, and the Kinshasa Reactor: Implications of Three Case Studies for Combating Nuclear Terrorism*, by Sara Daly, John Parachini, and William Rosenau (Santa Monica, Calif.: Rand Corp., 2005); "The Changing Proliferation Threat," by John F. Sopko, in *Foreign Policy*, Winter 1996–1997; and "The AFP Investigation into Japanese Cult Activities in Western Australia," a case study compiled by Richard Crothers of the Australian Federal Police, Apr. 24, 2007. The psychology and doctrine of the group is dissected in *Destroying the World to Save It: Aum Shinrikyo, Apocalyptic Violence, and the New Global Terrorism*, by Robert Jay Lifton (New York: Owl Books, 1998), and *The End of Time: Faith and Fear in the Shadow of the Millennium*, by Damian Thompson (Hanover, N.H.: University Press of New England, 1996). The rich history of apocalyptic thinking through history is explored in *Cosmos, Chaos, and the World to Come: The Ancient Roots of Apocalyptic Faith*, by Norman Cohn (New Haven, Conn.: Yale University Press, 1993); *Apocalypses: Prophesies, Cults, and Millennial Beliefs Through the Ages*, by Eugen Weber (Cambridge, Mass.: Harvard University Press, 1999); *Apocalypse: On the Psychology of Fundamentalism in America*, by Charles B. Strozier (Boston: Beacon Press, 1994); and *When Time Shall Be No More*, by Paul S. Boyer (Cambridge, Mass.: Belknap, 1992).

CHAPTER 5: TWO RUSHES

The buying policies of the Atomic Energy Commission and the widespread social effects of uranium mining on the Colorado Plateau are examined in *Yellowcake Towns: Uranium Mining Communities in the American West*, by Michael Amundson (Boulder: University of Colorado Press, 2002), and *Quest for the Golden Circle: The Four Corners and the Metropolitan West, 1945–1970*, by Arthur R. Gomez (Albuquerque: University of New Mexico Press, 1994). Also of great help was the dissertation "A History of the Uranium Industry on the Colorado Plateau," by Gary Shumway, University of Southern California, Jan. 1970. Shumway's master's thesis, "The Development of the Uranium Industry in San Juan County, Utah," Brigham Young University, July 1964, also provided source material. Some background on the geology of the Colorado Plateau and the initial mining bonanza was taken from "The Uranium Rush," by Tom Zoellner, the *American Heritage of Invention & Technology*, Summer 2000; *The Redrock Chronicles*, by Tom H. Watkins (Baltimore: Johns Hopkins University Press, 2000); and "The Time of the Great Fever," by Larry Meyer, *American Heritage*, June/July 1981. The "Good for nothing" quote originally appeared in *Grand Memories*, by Phyllis Cortes (Grand County, Utah: Daughters of the Utah Pioneers, 1978), and was quoted in "Hot Rocks Made Big Waves," by Amberly Knight, in *Utah Historical Quarterly*, Winter 2001, as well as the Fall 2006 edition of *Canyon Legacy*, the journal of the Dan O'Laurie Museum in Moab. The hyperbolic *True West* article was reprinted in *Canyon Legacy*, Summer 2006. The bogus uranium foundations were first reported in "Quack-

ery in the Atomic Age," in *BusinessWeek*, Aug. 29, 1953. Some of the quotes
from Utah miners in the 1950s—including Jerry Anderson and Oren Zufelt—
were drawn from an archive of oral history at California State University, Ful-
lerton, established by Professor Gary Shumway. The naming controversy in
Moab comes from *The Far Country: A Regional History of Moab and La Sal,
Utah*, by Faun McConkie Tanner (Salt Lake City: Olympic Publishing, 1976).
The anecdotes about the circling airplanes and the staked highway are from *One
Man's West*, by David Lavender (New York: Doubleday, 1964). Other details
came from "Book 61: The Moab, Utah Story," by Clement K. Chase, a personal
memoir dated December 1997, as well as *Charlie Steen's Mi Vida*, by Maxine
Newell (Moab, Utah: Moab's Printing Place, 1992). Quotes from Paddy Marti-
nez were taken from "The Coming Thing," by Daniel Lang, in the *New Yorker*,
Mar. 21, 1953. The Popeye cartoon was described in "Uranium on the Cranium,"
an essay by Michael Admundson in the anthology *Atomic Culture: How We
Learned to Stop Worrying and Love the Bomb*, ed. Admundson and Scott C.
Zeman (Boulder: University Press of Colorado, 2005). Some details on the color-
ful life of Charlie Steen were drawn from "Uranium Millionaire," by Jack Good-
man, the *New York Times*, Oct. 17, 1954; "Uranium Mining Stocks Feed
Gambling Fever," by Jack R. Ryan, the *New York Times*, June 20, 1954; "Uranium: Jackpot in Utah," in *BusinessWeek*, Aug. 1, 1953; "Ordinary Was Radioactive to Charlie Steen, the Uranium King," by Gary Massaro, *Rocky Mountain
News*, Mar. 16, 2006; and "Fallout in the Family," by Ward Havarky, in Denver's
Westword, Feb. 19, 1998. But this section depended most heavily upon *Uranium
Frenzy*, by Raye Ringholz (Albuquerque: University of New Mexico Press,
1989), the definitive history of the uranium story in the American Southwest,
which also provides the most comprehensive account available of Steen's Mi Vida
strike and the efforts of the AEC to cover up the radioactive disaster. The angry
speech at Texas Western College, as well as several other episodes in the life of
Steen, and of Utah's stock bubble, were drawn from Ringholz's impressive work.
John Black's remembrances were in *Blue Mountain Shadows*, Winter 2001,
published by the San Juan County Historic Society. Tom McCourt's recollec-
tions of Moab, as well as his reflections on patriotism, were taken from *White
Canyon: Remembering the Little Town at the Bottom of Lake Powell* (Price,
Utah: Southpaw Publications, 2003). Joe Blosser's quote is in "Uranium Is
People," by Paul Schubert, in *Empire* magazine, reprinted in *Reader's Digest*,
Mar. 1953. The Dr. Seuss incident is related in "Man on Probation in Attempt
to Extort Dr. Seuss Estate," by Onell R. Soto, in the *San Diego Union Tribune*,
Aug. 21, 2004. Background on the AEC cover-up was drawn from *The Myths of*

August, by Stewart Udall (New York: Pantheon, 1994). The Navajo section drew from the excellent *If You Poison Us: Uranium and Native Americans,* by Peter Eichstaedt (Santa Fe, N.M.: Red Crane Books, 1994), which contains details on the economics of reservation mining, as well as interviews with miners, including Ben Jones, and information on hazards that still remain. "Blighted Homeland," a four-part series by Judy Pasternak in the *Los Angeles Times,* Nov. 19–22, 2007, examined the lasting health and environmental impacts and was the source of the "Saudi Arabia" remark. The story "Udall: Navajo 'Cancer Free' Until Uranium," by Kathy Helms, in the *Gallup Independent,* Nov. 15, 2007, had further historical background. The comments of the Navajo miner Willie Johnson came from "Toxic Targets," by Jim Motavalli, in *E* magazine, July 1998. An evocative fictional account of the afterlife of a uranium town can be found in the novel *Yellowcake,* by Ann Cummins (Boston: Houghton Mifflin, 2007). Health data comes from "Lung Cancer in a Nonsmoking Uranium Miner," by Karen B. Mulloy et al., in *Environmental Health Perspectives* 109 (2001); "The History of Uranium Mining and the Navajo People," by Doug Brugge and Rob Goble, in *American Journal of Public Health,* Sept. 2002; "Lung Cancer Risk Among German Male Uranium Miners," by B. Grosche et al., in the *British Journal of Cancer,* Oct. 2006; and "Diseases of Uranium Miners and Other Underground Miners Exposed to Radon," by J. M. Samet and D. W. Mapel, in *Environmental and Occupational Medicine,* 1998. The foremost historians of the postwar uranium era in Europe are Rainer Karlsch of the Free University of Berlin and Zbynek Zeman of Oxford University. Their recently translated book *Uranium Matters: Central European Uranium in International Politics, 1900–1960* (Budapest: Central European University Press, 2008), was a source for a few details in this chapter, including the supposed "acts of sabotage" at Wismut and the bedsheets at St. Joachimsthal. Details on the diplomatic background to the Czechoslovakian-Soviet accords, as well as a wealth of information about the Jachymov gulag, come from two illuminating articles by Zbynek Zeman: "The Beginnings of National Enterprise Jachymov (Joachimsthal)," in *Der Anschnitt,* Mar. 1998, and "Czech Uranium and Stalin's Bomb," in *Historian,* Autumn 2000. Frantisek Sedivy was kind enough to give me a copy of his autobiography, *The Legion of the Living* (Prague: Eva-Milan Nevole, 2003), which supplied some of the details of his incarceration. He also wrote a novel based on his experience called *Under the Tower of Death* (Prague: Eva-Milan Nevole, 2003). Translations of key portions of these important books were supplied to me by Marketa Naylor. The anecdote about the dog and the description of the town in St. Joachimsthal comes from "Soft Norms in a Spa," by Joseph Wechsberg, in the *New Yorker*, May 3, 1952. Quotes from Joseph Stalin were in *Stalin and the Bomb*, by David Holloway (New Haven, Conn.: Yale University Press, 1994), and also cited in *Bomb Scare*, by Joseph Cirincione (New York: Columbia University Press, 2007). Using previously sealed documents from the Eastern bloc, Norman Naimark pieced together an account of the labor situation at Wismut for a section of his book *The Russians in Germany: A History of the Soviet Zone of Occupation, 1945–49* (Cambridge, Mass.: Belknap Press, 1995), from which I garnered a few details. Information about the culture and operations of Wismut were drawn from materials on exhibit at the excellent Museum Uranbergau in Schlema, Germany; I received translation assistance from the curator, Herman Meinel, and from Martha Brantley, who helped me speak with former Wismut miners. Some health statistics were taken from "Wismut: Uranium Exposure Revisited," by Heinz Otten and Horst Schulz, a research paper published by the International Labor Organization (New York: United Nations, 1998). The former Wismut manager Nikolai Grishin disclosed some state secrets in "The Saxony Mining Operation," reprinted in *Soviet Economic Policy in Postwar Germany* (New York: Research Program on the USSR, 1953). The extent of Western intelligence knowledge about Wismut, though colored by some period bias, can be found in the background paper "Zhukov and the Atomic Bomb," author unknown (Munich: Radio Free Europe, 1957), as well as *Forced Labor in the "People's Democracies,"* by Richard K. Carlton (Munich: Free Europe Committee, 1955). The policies and practices at Wismut were examined in the unpublished doctoral dissertation "The Quest for Uranium: The Soviet Uranium Mining Industry in Eastern Germany, 1945–1967," by Traci Heitschmidt, on file in the library at the University of California–Santa Barbara. The anecdote about the hidden Soviet enrichment plants comes from *Shadow Flights: America's Secret Air War Against the Soviet Union*, by Curtis Peebles (New York: Presidio Press, 2000), and the declassified background paper titled "On the Soviet Nuclear Scent," by Henry S. Lowenhaupt of the Central Intelligence Agency (undated). Some information about the gulag period is drawn from *Jachymov: The City of Silver, Radium, and Therapeutic Water*, by Hana Hornatova (Prague: Medeia Bohemia, 2000). The discovery stories were recounted in "Miner's Luck," by Henry Winfred Splitter, in *Western Folklore*, Oct. 1956, as well as *History of California*, by Theodore Henry Hittell (San Francisco: N. J. Stone, 1898).

CHAPTER 6: THE RAINBOW SERPENT

Some of the color of the early days of Australia's uranium rush comes from *The Uranium Hunters*, by Ross Annabell (Adelaide, Australia: Rigby, 1971). The petroglyph of the Rainbow Serpent is discussed in "Flood Gave Birth to World's Oldest Religion," by Leigh Dayton, in *New Scientist*, Nov. 11, 1996. This section could not have been written without the help of Joe Fisher, who donated a scrupulously documented account of his life: a self-published autobiography in two volumes. These valuable books are *Trials and Triumphs in the Northern Territory and Northern Australia: From Cape York to the Kimberleys, 1954–2002* (Melbourne, Australia: S. R. Frankland, 2002) and *Battles in the Bush: The Batavia Goldfields of Cape York* (Melbourne, Australia: S. R. Frankland, 1998). Political background on the uranium debate comes partially from "The Rise of Anti-Uranium Protest in Australia," by Sigrid McCausland, a paper submitted to the Australasian Political Studies Association conference, Oct. 2000; *Uranium on Trial*, by Stuart Butler, Robert Raymond, and Charles Watson Moore (Sydney: New Century Press, 1977); the newspaper article "Kakadu: 'Scruffy and a Bore,'" in the *Northern Territory News*, Aug. 8, 1978, reprinted in Fisher, *Trials and Triumphs* and *Battles in the Bush*; "Yellowcake Country: Australia's Uranium Industry," a paper prepared in 2006 by Beyond Nuclear Initiative, Melbourne, Australia; and "Nuclear Power No Solution to Climate Change," a paper prepared for Friends of the Earth et al., Sydney, Sept. 2005. The history of the town of Jabiru is covered in *Yellowcake and Crocodiles*, by John Lea and Robert Zehner (Sydney: Allen & Unwin, 1986). The opinionated and well-written memoir *Jabiluka: The Battle to Mine Australia's Uranium*, by Tony Grey (Melbourne, Australia: Text Publishing, 1994), provided key context and details, as did *Kakadu: The Making of a National Park*, by David Lawrence (Melbourne, Australia: Melbourne University Publishing, 2000). "ERA 2005 Annual Report," prepared by Energy Resources of Australia, has some details about the operation of the Ranger Mine. An impressive amount of original research on the Uranium Club can be found in *Yellowcake*, by J. Taylor and Michael Yokell (Amsterdam: Elsevier, 1979). A layman's guide to their activities is in *The Politics of Uranium*, by Norman Moss (New York: Universe Publishers, 1984). A contemporaneous report of the club's later days is in "'It Worked for the Arabs . . .'" in *Forbes*, Jan. 15, 1975. Some of the lore of Rio Tinto was drawn from *The Cooperative Edge: The Internal Politics of International Cartels*, by Debora L. Spar (Ithaca, N.Y.: Cornell University Press, 1994). Sir Val Duncan's imperious quote was recalled in "A Very British Coup" in the *Daily Mail*, Mar.

13, 2006. Some continental history was taken from *Aboriginal Australians: Black Responses to White Dominance, 1788–2001,* by Richard Broome (Sydney: Allen & Unwin, 2002); *Dreamings: The Art of Aboriginal Australia,* by Peter Sutton (New York: George Braziller, 1988); and *Arguments About Aboriginals,* by L. R. Hiatt (Cambridge: Cambridge University Press, 1996). This section also drew from newspaper accounts of the Jabiluka protests in the *Age, Sydney Morning Herald,* and *Northern Territorial News.*

CHAPTER 7: INSTABILITY

Statistics and history on the Areva mines in Niger were drawn from a paper read at the 2004 annual symposium of the World Nuclear Association in London: "Uranium Mining in Niger: Status and Perspectives of a Top Five Producing Country," by George Capus, Pascal Bourrelier, and Moussa Souley. "Country Report," by the Economist Intelligence Unit, Feb. 2, 2007, was also helpful, as was the article "Niger: Uranium—Blessing or Curse?," by the United Nations Office for the Coordination of Humanitarian Affairs, Oct. 10, 2007. Accounts of the violence in Arlit were taken from the news dispatches "France Sees Areva Progress, Offers Niger Mine Aid," by Abdoulaye Massalatchi of Reuters, Aug. 4, 2007; "Niger's Uranium Industry Threatened by Rebels," by Andrew McGregor, in *Terrorism Focus,* July 31, 2007; "Niger Rebels Pressure Uranium Mines," by James Finch, in *Stock Interview,* July 9, 2007; "Uranium Worth a Fight, Niger Rebels Say," by Tristan McConnell, in the *Christian Science Monitor,* Oct. 21, 2007; and "Five Wounded as Bus Hits Landmine in Niger," by the South Africa Press Association and Agence France-Presse, Nov. 23, 2007. The United Nations Office for the Coordination of Humanitarian Affairs also produced two relevant bulletins: "New Tuareg Rebel Group Speaks Out," May 12, 2007, and "Five Killed as Army Clashes with Tuaregs in Desert North," Oct. 7, 2007. The account of Rocco Martino's dealings with Elisabetta Burba comes from the well-reported book *The Italian Letter: How the Bush Administration Used a Fake Letter to Build the Case for War in Iraq,* by Peter Eisner and Knut Royce (New York: Rodale, 2007). Eisner offers a compressed version in "How a Bogus Letter Became a Case for War," in the *Washington Post,* Apr. 3, 2007. I also drew from other accounts for this section, including "The Italian Job: How Fake Iraq Memos Tripped Up Ex-Spy," by Jay Solomon and Gabriel Kahn, in the *Wall Street Journal,* Feb. 22, 2006; "The War They Wanted, the Lies They Needed," by Craig Unger, in *Vanity Fair,* July 2006; and "The Italian Job," by Laura Rozen, in the *American Prospect,* Mar. 2006. The rehashing of the U.S. buildup to war was partly drawn from Unger, "The War They Wanted," and Eisner and Royce, *The Italian Letter*, as well as "U.S. Claim on Iraqi Nuclear Program Is Called into Question," by Joby Warrick, in the *Washington Post*, Jan. 24, 2003, and "What I Didn't Find in Africa," by Joseph Wilson, in the *New York Times*, July 6, 2003. Reflections on nuclearism were taken from the lecture "The Image of the End of the World: A Psychosocial History," by Robert Jay Lifton, at Salve Regina College in Newport, R.I., in 1983, and reprinted in *Facing Apocalypse* (Dallas: Spring Publications, 1987). Thoughts on the state of Islamic science came from "Myth-Building: The 'Islamic' Bomb," by Pervez Hoodbhoy, in the *Bulletin of the Atomic Scientists*, June 1993; "Islam and Science—Unhappy Bedfellows," by Pervez Hoodbhoy, in *Global Agenda*, Jan. 2006; *The Arab Human Development Report 2002: Creating Opportunities for Future Generations*, by the United Nations Development Program and the Arab Fund for Economic and Social Development (also quoted by Hoodbhoy); and *What Went Wrong?*, by Bernard Lewis (Oxford: Oxford University Press, 2002). Al-Qaeda's history with uranium—as well as the economics of homemade atomic bombs—is partly covered in "The Bomb in the Backyard," by Peter D. Zimmerman and Jeffrey G. Lewis, in the *National Post* of Canada, Dec. 20, 2006, as well as *The Looming Tower*, by Lawrence Wright (New York: Random House, 2006). Some biographical information about Mahmoud Ahmadinejad is in the newspaper stories "Waiting for the Rapture in Iran," by Scott Peterson, in the *Christian Science Monitor*, Dec. 21, 2005; "'Divine Mission' Driving Iran's New Leader," by Anton La Guardia, the *Telegraph*, Jan. 15, 2006; and "Nuclear Armed Iran Risks World War, Bush Says," by Sheryl Gay Stolberg, in the *New York Times*, Oct. 18, 2007. Details of the experimental centrifuges at Natanz were drawn from "A Tantalizing Look at Iran's Nuclear Program," by William J. Broad, in the *New York Times*, Apr. 29, 2008. Iran's history and attitude toward nuclear science can be found in the background paper "Iran: Nuclear Chronology," by the staff of the Monterey Institute for International Studies, 2003; and the book *The Nuclear Sphinx of Tehran: Mahmoud Ahmadinejad and the State of Iran*, by Yossi Melman and Meir Javedanfar (New York: Carroll & Graf, 2007). Also, the news stories "Rafsanjani: Iran Will Get Its Nuclear Rights with Wisdom," by the Islamic Republic News Agency, Jan. 11, 2006; "Rafsanjani: Europe Indebted to Muslims for Scientific Advancement," by the Islamic Republic News Agency, Aug. 15, 2007; "Iran Admits Nuclear Secrecy," by the Associated Press, Mar. 7, 2005; "Western Pressure Irks Average Iranians," by Angus McDowell, the *Christian Science Monitor*, Apr. 24, 2006; "Across Iran, Nuclear Power Is a Matter of Pride," by Neil MacFarquhar, the *New York Times*, May 29, 2005;

"Iran Looks to Science as Source of Pride," by Anne Barnard, the *Boston Globe*, Aug. 22, 2006; "The Riddle of Iran," in the *Economist*, July 21, 2007; and "Satellite Images Show Work Near Iran Nuclear Site," by Reuters, reprinted in Istanbul's *Today's Zeman*, July 11, 2007. The section on the logistics of uranium acquisition was drawn from "Preventing Nuclear Terrorism: Reducing the Danger of Highly Enriched Uranium," a 2003 paper by Hui Zhang of the Kennedy School of Government at Harvard University; Zimmerman and Lewis, "The Bomb in the Backyard"; "Stockpiles Still Growing," by David Albright and Kimberly Kramer, in *Bulletin of the Atomic Scientists*, Nov./Dec. 2004; "Nuclear Fuel Is Widespread," by Sam Roe, in the *Chicago Tribune*, Feb. 4, 2007; and especially "Eliminating Excessive Stocks of Highly Enriched Uranium," by Morten Bremer Maerli and Lars van Dassen, in *Pugwash Issue Brief*, published by the Council of the Pugwash Conferences on Science and World Affairs, Apr. 2005. The quote from Ashton Carter is in "Responding to Iran's Nuclear Ambitions: Next Steps," the transcript of a Sept. 19, 2006, hearing before the U.S. Senate Committee on Foreign Relations. William Langewiesche envisions a uranium-theft scenario and also pays an eye-opening visit to a checkpoint on the Georgian border with Armenia in "How to Build an Atomic Bomb," in the Dec. 2006 issue of the *Atlantic*, reprinted in a revised version in *The Atomic Bazaar*, by William Langewiesche (New York: Farrar, Straus and Giroux, 2007). The Logan Airport incident was mentioned in John McPhee's *The Curve of Binding Energy* (New York: Farrar, Straus and Giroux, 1973), and the Erwin safety record was examined in *Nukespeak: The Selling of Nuclear Technology in America*, by Stephen Hilgartner, Richard Bell, and Rory O'Conner (San Francisco: Sierra Club Books, 1982). The incident in Dalhart was reported in the Sept. 22, 1951, United Press dispatch "'Plaything' of Three Boys Turns Out to Be Uranium," and also in "Buried Treasure," in *Time*, Oct. 1, 1951. The story of Sanford Simons was recalled in *Doomsday Men*, by P. D. Smith (London: Allen Lane, 2007). The Shanghai black market was detailed in a declassified memo of June 24, 1946, from the Strategic Services Unit of the War Department entitled "Ramona (Summary Report for June)." Colombia's uranium seizure was told of in the National Public Radio report "Colombia Reflects Rising Threat of Nuclear Terrorism," by Tom Gjelten, and broadcast Apr. 21, 2008. Further information on loose uranium was drawn from "Czech Seize Migrating Uranium," by Mark Hibbs, in *Bulletin of the Atomic Scientists*, Mar./Apr. 1993; "Nuclear Cleanup's Trudge," by David E. Hoffman, in the *Wall Street Journal*, Aug. 31, 2007; "At Mayak, Lax Security Worries U.S.," by Ann Imse, *Rocky Mountain News*, Feb. 22, 2003; the research paper "Recent Weapons-Grade Uranium Smuggling

Case: Nuclear Materials Are Still on the Loose," by Elna Sokova, William C. Potter, and Christina Chuen, Jan. 26, 2007, published by the Monterey Institute of International Studies; and "Nuclear Smuggling, Rogue States, and Terrorism," by Rensselaer Lee, in *The China and Eurasia Forum Quarterly*, Apr. 2006. Spying and asbestos at the IAEA were discussed in Melman and Javedanfar, *The Nuclear Sphinx of Tehran*. The bale-of-marijuana aphorism is related in "The Seven Myths of Nuclear Terrorism," by Matthew Bunn and Anthony Wier, *Current History*, Apr. 2005, and is also cited in Langewiesche, *The Atomic Bazaar*. Bunn's quote about poor oversight in Russia comes from his *Securing the Bomb 2007* (Cambridge, Mass., and Washington, D.C.: Project on Managing the Atom, Harvard University, and Nuclear Threat Initiative, Sept. 2007). The movement of nonnuclear goods over the Georgian border is examined in the 2004 research paper "Smuggling Through Abkhazia and Tskhinvali Region of Georgia," by Alexandre Kukhianidze, Alexandre Kupatadze, and Roman Gotsiridze and published by the Transnational Crime and Corruption Center in Tbilisi. The tale of Oleg Khinsagov was first broken in the story "Smuggler's Plot Highlights Fear Over Uranium," by Lawrence Scott Sheets and William J. Broad, in the *New York Times*, Jan. 25, 2007, from which I drew some details. Further details were from "A Smuggler's Story," by Lawrence Scott Sheets, in the *Atlantic*, Apr. 2008. Some history of the Darial Gorge is in *The Land of the Czar*, by O. W. Wahl (London: Chapman and Hall, 1875).

CHAPTER 8: RENAISSANCE

General overviews of supply and demand are contained in the reports *The Global Nuclear Fuel Market*, by the staff of the World Nuclear Association, 2005, and *Investing in the Great Uranium Bull Market: A Practical Investor's Guide to Uranium Stocks* (Sarasota, Fla.: Stock Interview, 2006). Further details are in "Atomic Renaissance," in the *Economist*, Sept. 8, 2007; "The New Economics of Nuclear Power," by the staff of the World Nuclear Association, 2005; "A Rush for Uranium," by Susan Moran and Anne Raup, in the *New York Times*, Mar. 28, 2007; "Nuclear Power: Winds of Change," by Michael Campbell et al., a paper from the American Association of Petroleum Geologists, Mar. 31, 2007; and "Solving 'Fission Impossible,'" by Daniel Gross, in *Newsweek*, Oct. 29, 2007. The debate about the climate benefits from nuclear energy are reflected in the following: "Nuclear Power Is the Only Green Solution," by James Lovelock, in the London *Independent*, May 24, 2006; the position paper "Environmentalists Do Not Support Nuclear Power," by Jim Green, published by Friends of the Earth,

Australia, May 11, 2007; "Atomic Myths, Radioactive Realities: Why Nuclear Power Is the Poor Way to Meet Energy Needs," by Arjun Makhijani in the *Journal of Land, Resources & Environmental Law* 24, no. 1 (2004); and "Pelosi Reconsiders Nuclear Power," in the *Wall Street Journal,* Feb. 8, 2007. Statistics and background on Chinese coal came from two *New York Times* stories: "Dangerous Coal Mines Take Human Toll in China," by Erik Eckholm, June 19, 2000, and "Pollution from Chinese Coal Casts a Global Shadow," by Keith Bradsher and David Barboza, June 11, 2006. Details on Senator Pete Domenici's lobbying on behalf of the nuclear industry and his statements at the Eunice, New Mexico, ground-breaking are in the well-reported Jan. 2007 package of stories "Power Play: New Dawn for Nuclear Energy?," by Mike Stuckey and John W. Schoen, on MSNBC.com. Some background on the decision is in "Waste Issues Dog Uranium Plant Build," by Ben Neary, in the *Santa Fe New Mexican,* Dec. 9, 2003; "Recent Almelo Visitors Speak to Eunice Rotary Club," in the *Eunice News,* Dec. 13, 2007; "Texas Senate Approves Fee to Bury Nuclear Waste in Andrews," by John Reynolds, in the *Lubbock Avalanche-Journal,* May 5, 2005; and "Dangerous Liaisons," by Marilyn Berlin Snell, in *Sierra,* May/June 2005. General background on the region is in *"Little Texas": Beginnings in Southeastern New Mexico,* by May Price Mosley (Roswell, N.M.: Hall-Poorbaugh Press, 1973). Paducah's early history is recounted in the special section "The Atomic Plant's 40th Anniversary," reprinted in the *Paducha Sun,* Nov. 3, 1992. The less happy environmental aftereffects are detailed in two stories by Joby Warrick in the *Washington Post:* "Paducah Plant Spewed Plutonium," Oct. 1, 2000, and "Nuclear Bomb Risk Revealed at Kentucky Uranium Plant," Feb. 11, 2000. Information on the Piketon plant came from the press release "USEC Will Fuel Nuclear Revival, CEO Tells Shareholders," Apr. 25, 2006, and the newspaper story "Costly Centrifuge Plan Key to Piketon Revival," by Tom Beyerlein and Lynn Hulsey, in the *Dayton Daily News,* Nov. 14, 2006. Some recent developments on the Arizona Strip are in "Power Surge," by Max Jarman, in the *Arizona Republic,* May 28, 2006. Fragments from the life of Bob Adams and the history of his Energy Fuels Nuclear company are told in "Bob Adams: Positive Energy Force in the Yampa Valley," by Rod Hanna, in *Steamboat Springs,* Summer 1980; "Bob Adams: 1917–1982," in the *Steamboat Pilot,* Sept. 30, 1982; "Home on the Range No More: The Boom and Bust of a Wyoming Uranium Mining Town, 1957–1988," by Michael A. Amundson, in the *Western Historical Quarterly,* Winter 1995; and *Quest for the Pillar of Gold: The Mines and Miners of the Grand Canyon,* by George E. Billingsley, Earle E. Spamer, and Dove Menkes (Grand Canyon Village, Ariz.: Grand Canyon Association, 1987). The colorful early history of the Vancouver exchange, including the Pine Point anecdote, comes from *Fleecing the Lamb: The Inside Story of the Vancouver Stock Exchange*, by David Cruise and Allison Griffiths (Vancouver: Douglas & McIntyre, 1987). Further information was drawn from "The Scam Capital of the World," by Joe Queenan, in *Forbes*, May 29, 1989; "Salt for the Bre-X Wounds," in *Macleans*, Mar. 2, 1998; "The Ghost of Bre-X Rises," by Steve Maich, in *Macleans*, June 13, 2005; "Geologists Still Have Something to Answer For," by David Baines, in the *Vancouver Sun*, Aug. 18, 2007; and "U.S. Gets Burned by Lax Canadian Oversight," by Robert McClure, in the *Seattle Post-Intelligencer*, June 13, 2001. The broker's quote comes from the Oct. 28, 1997, Reuters dispatch "Bre-X Joins Forces with Suharto's Son, Stock Soars," by Heather Scoffield, and quoted in the July 31, 2007, court judgment *Her Majesty the Queen v. John Bernard Felderhof*, by Justice Peter Hyrn in the Ontario Justice Court. General background on Mongolia is drawn from *In the Empire of Genghis Khan*, by Stanley Stewart (Guilford, Conn.: Lyons Press, 2004); some surprising aspects of Khan's reign are in "To the Left of Chinggis Khan," by Timothy May, in *World History Connected*, Nov. 2006. Development schemes for Mardai are disclosed in "Western Prospector Builds on Soviet-Era Uranium Project," by Stephen Stakiw, in the *Northern Miner*, Nov. 25, 2005, and the corporate report "Gurvanbulag Uranium Mine and Mill Development Plans," by Emeelt Mines LLC, June 2007. Ivanhoe's recent history is in "The New El Dorado," by Michael Schuman, in *Time International*, Aug. 7, 2006; "Your Risk, His Reward," by David Baines, *Canadian Business*, June 1997; and "Big Dig: Mongolia Is Roiled by Miner's Huge Plans," by Patrick Barta, in the *Wall Street Journal*, Jan. 4, 2007.

CHAPTER 9: LEGACY

Gerard Holden's quotes came from the Brinkley Mining announcement "Agreement in DRC," dated July 11, 2007, and also the Sept. 18, 2007, news dispatch "Brinkley Hits Back in DRC Uranium Fracas," by Allan Seccombe of Miningmx .com. Further background is in "Congo Purge Puts Brinkley Deal in Doubt," by Ben Laurence, in the *Sunday Times*, Sept. 16, 2007; the British Broadcasting Company story "DR Congo 'Uranium Ring Smashed,'" Mar. 8, 2007; the Reuters dispatch "Congo Keeps Uranium Riches Under Wraps," Dec. 10, 2007; and the article "Uranium Smuggling Allegations Raise Questions Concerning Nuclear Security in the Democratic Republic of Congo," by Peter Crail and Johan Bergenas, in *WMD Insights*, Apr. 2007. Some reactor information comes from "Nuclear Technical Cooperation: A Right or a Privilege?," by Jack Boureston and Jennifer Lacey, in *Arms Control Today*, Sept. 2007. The Yucca Mountain plan has felled enough trees to fill up its repository with paper products alone. A good general primer to the complicated scenario at Yucca Mountain, as well as information on the dilemma of the Skull Valley Band of Goshute, is *Nuclear Waste Stalemate: Political and Scientific Controversies*, by Robert Vandenbosch and Susanne E. Vandenbosch (Salt Lake City: University of Utah Press, 2007). Designs for a permanent nuclear waste marker are discussed in *Expert Judgment on Markers to Deter Inadvertent Human Intrusion Into the Waste Isolation Pilot Plant*, by Kathleen Trauth, Stephen Hora, and Robert Guzowski (Albuquerque, N.M.: Sandia National Laboratory, 1993). News coverage of the plan was in "An Alert Unlike Any Other" by Charles Piller in the *Los Angeles Times*, May 3, 2006; and "Early Warning: How to Alert Earthlings of Yucca's Waste" by Peter Waldman in the *Wall Street Journal*, February 10, 2003. The history and flora of the region are in "Reading the Stones: The Archeology of Yucca Mountain" by William T. Hartwell and David Valentine, (Las Vegas: Desert Research Institute, 2009). Migrations of political fauna are seen in "Mighty Expensive Dinosaur" in the *Las Vegas Sun*, June 8, 2009; "The 'Screw Nevada Bill' and How it Stymied U.S. Nuclear Waste Policy" by John J. Fialka, in the *New York Times*, May 11, 2007; and "This is the Place for Waste," by Brent Israelsen in the *Salt Lake Tribune*, Sept. 9, 2002. Thoughts on the Nevada desert are drawn from *The Mojave* by David Darlington (New York: Henry Holt, 1996); *Nevada: The Great Rotten Borough 1859–1964* by Gilman M. Ostrander (New York: Knopf, 1966); and *In Nevada: The Land, The People, God, and Chance* by David Thomson (New York: Knopf, 1999). Mark Twain's unfortunate diatribe against the Goshute is in *Roughing It* (New York: American Publishing Company, 1871). The ugly history of Love Canal and the army's role in dumping atomic-era trash there can be found in the damning two-volume investigation delivered to the New York State Assembly by the Task Force on Toxic Substances on January 29, 1981. The report is entitled *The Federal Connection: A History of U.S. Military Involvement in the Toxic Contamination of Love Canal and the Niagara Frontier Region*. Other valuable sources included *Inventing Niagara: Beauty, Power, and Lies*, by Ginger Strand (Simon and Shuster, 2008); and *High Hopes: The Rise and Decline of Buffalo, New York*, by Mark Goldman (Albany: State University of New York Press, 1983).

Estimates of the Russian HEU stockpile are in *Plutonium and Highly Enriched Uranium 1996: World Inventories, Capabilities, and Policies,* by David Albright, Frans Berkhout, and William Walker (New York: Oxford University Press, 1997), and quoted in "Russia: Fissile Material and Disposition," by the Center for Nonproliferation Studies. The controversial Department of Energy survey is called *HEU: Striking a Balance—A Historical Report on the United States Highly Enriched Uranium Production, Acquisition, and Utilization Activities from 1945 Through September 30, 1996.* The circumstances of its suppression and eventual release are discussed in the article "The U.S. Highly Enriched Uranium Declaration: Transparency Deferred but Not Denied," by Steven Aftergood and Frank N. von Hippel, in *Nonproliferation Review,* Mar. 2007. The phenomenon at Oklo was explained by George A. Cowan in *Scientific American,* July 1976. The street address for the site of the Archer Daniels Midland warehouse—2377 Richmond Terrace on Staten Island, right under the Bayonne Bridge—was drawn from the diligent work of Timothy L. Karpin and James M. Maroncelli, *The Traveler's Guide to Nuclear Weapons Sites* (Lacey, Wash.: Historical Odysseys Publishers, 2002), supplied to me by Robert S. Norris.

INDEX